THE FUTURE OF FOOD AND AGRICULTURE AFTER COVID

Disruptive Trends and Technology That Will Shape the Post-Pandemic World

JASON SCHENKER

THE FUTURE OF FOOD AND AGRICULTURE AFTER COVID

Disruptive Trends and Technology That Will Shape the Post-Pandemic World

BY JASON SCHENKER

ISBN: 978-1-946197-66-5 *Paperback*
978-1-946197-67-2 *Ebook*

PRESTIGE
PROFESSIONAL PUBLISHING

For my loving wife, Ashley.

CONTENTS

CONTENTS

CONTENTS

PREFACE

LOOKING AHEAD AFTER COVID

Food and agriculture concerns have been at the heart of the COVID-19 pandemic, economic shutdown, and recession. As the entire world was shoved into the lowest level of Maslow's hierarchy of needs, fears about food supplies became rampant.

In April 2020, my book *The Future After COVID* was released. In it, I discussed the importance of food and agriculture at a time when uncertainty had risen significantly.

In the end, technology helped save the day during the pandemic shutdown, as e-commerce and food delivery became critical for keeping people safe, the economy intact, and society stable. Looking ahead, food security and stable agricultural activities will become increasingly important priorities across economies.

The backdrop of the ongoing U.S.-China trade conflict has hinged on agricultural purchases, and it underscores the risks to global food and agricultural trade going forward. Furthermore, a long-term rise in global population and global wealth will drive calorie and protein needs beyond what the world currently provides.

This is where technology will again need to provide critical solutions. Global needs and the prospects of global conflict will drive critical trends in food and agriculture. Sustainability, shortages, water crises, and other needs will come to a head in the decade ahead and beyond.

These trends and the technologies that will disrupt, impact, and resolve some of them will affect many people in the decade following the outbreak of the COVID-19 pandemic. That is the big idea of this book; to evaluate what comes next. It's why the book is titled *The Future of Food and Agriculture After COVID: Disruptive Trends and Technology That Will Shape the Post-Pandemic World.*

Acknowledgements

This book represents an attempt to put the present in context and examine what comes next for food and agriculture in the decade ahead and beyond. But no book is done completely alone.

There is editorial work, file conversion, design, and project management. And those tasks require a team. Along those lines, I want to thank Prestige Economics, The Futurist Institute, and the Prestige Professional Publishing staff for making this book a reality. Additionally, I want to especially thank **Nawfal Patel**, who managed the production of *The Future of Food and Agriculture After COVID.*

Most importantly, I want to thank my family for supporting me in my education, career, entrepreneurship, and authorship.

I am always most grateful for the support of my loving wife, **Ashley Schenker**, and to my wonderful parents, **Janet and Jeffrey Schenker**. My family supports me in countless ways by providing emotional support and editorial feedback.

Every time I write a book, it's a crazy experience that spills over into my family life, so to them and to everyone else who helped me in this process: Thank you!

Finally, thank you for buying this book. I hope you enjoy *The Future of Food and Agriculture After COVID.*

~ Jason Schenker
August 2020

INTRODUCTION

DISCUSSING THE FUTURE OF FOOD AND AGRICULTURE

Food and agriculture are parts of the economy that consumers have long taken for granted. With the onset of the COVID-19 pandemic, however, that changed drastically. But more changes lie ahead for how food is prioritized and delivered to consumers, and also for how agricultural challenges are met with technology in the future.

Some of the topics are unlikely to be surprising, including e-commerce and food delivery. But other technological topics are a bit more advanced and forward looking. While blockchain has already been used in some areas of agriculture and food supply chains, much greater use of blockchain in these areas is likely in the decade after COVID. Plus, more bleeding-edge technologies such as quantum computing are likely to become critical for addressing uncertainties in the field of agriculture.

In this book, I address this mix of changing dynamics, long-term trends, and deployment of new, emerging, and bleeding-edge technologies.

The Structure of This Book

In order to tackle the most important technology and trend factors in *The Future of Food and Agriculture After COVID* I have divided this book into seven main sections:

- **Critical Trends**
- **The Future of Calories**
- **Data Technologies**
- **Physical Technologies**
- **Sustainability and Shortages**
- **Disputed Trends**
- **Looking Ahead**

In the first section, **Critical Trends**, I discuss long-term demographic trends in place long before the COVID-19 pandemic, as well as some of the most important recent demographic and labor market dynamics. This includes a discussion on the past of agriculture (Chapter 2), the present of agriculture (Chapter 3), and the importance of demographic, economic, and technological drivers of the future of food and agriculture following the COVID-19 pandemic (Chapter 4).

When I examine demand and the **Future of Calories** in the second section of the book, I look at the most critical drivers on the demand side of food and agriculture. This includes the hunt for calories (Chapter 5) and the future of protein (Chapter 6). It also includes growing meat (Chapter 7) as well as hydroponics and the future of land (Chapter 8).

The third section of this book is dedicated to **Data Technologies**. This section includes an examination of critical growing and emerging data technologies, including e-commerce after COVID (Chapter 9), how to use data to unlock value, (Chapter 10), blockchain (Chapter 11), and quantum computing (Chapter 12).

In the fourth section of this book, **Physical Technologies**, I discuss the most important supply-side technologies including automation (Chapter 13) and operational drones (Chapter 14).

The fifth section of this book is dedicated to **Sustainability and Shortages**. This section addresses environmental, social, and governance (ESG) as well as agriculture issues like climate change and food deserts (Chapter 15). There's also a discussion of the move to "clean food" and the rise in demand for organic foods, halal food, and non–genetically modified food (Chapter 16). Plus, this section includes a discussion of waste to power (Chapter 17) and the prospects for the end of ethanol as a fuel, given rising global demand for food and calories in the decade ahead and beyond (Chapter 18).

The sixth section of this book is dedicated to **Disputed Trends** in agriculture. It includes a discussion of food extinction risks and GMOs (Chapter 19), the future of water (Chapter 20) as well a discussion about trade war dynamics, which present significant potential for disputes on a global scale with national security implications (Chapter 21). There is also an evaluation of the potential for acceptance and expansion of the marijuana business and its taxation (Chapter 22).

The seventh and final section of *The Future of Food and Agriculture After COVID* is titled **Looking Ahead**. This section includes future scenarios (Chapter 23) as well a conclusion to pull everything together (Chapter 24).

There are reasons to be hopeful for the future of food and agriculture after COVID, but there are also reasons to be concerned. It is my hope that the discussion in the chapters ahead will help you be prepared for the myriad changes coming to food and agriculture. After all, being forewarned is a critical prerequisite to being prepared for upside opportunities as well as downside risks.

CHAPTER 1

WHY I WROTE THIS BOOK

Although food security and shortages are persistent concerns and risks for many countries around the world at any point in time, it is not common for this to be a concern in the most developed OECD countries like the United States. But the COVID-19 pandemic, shutdown, and recession changed that.

Images of stores without paper products or with limited selections of food were reminiscent for many Americans, like myself, who lived through parts of the Cold War of the widespread images seen during the death throes of the U.S.S.R. Stories of people waiting in line for toilet paper and fresh food, as well as the images thereof, were widespread. They were further popularized in films like *Moscow on the Hudson*.

The image of the U.S.S.R. as one giant wasteland eclipsed what we today might think of as a food desert, where the availability of affordable fresh produce is limited, especially in urban areas — including in the United States. But the impact of COVID-19 resulted in widespread U.S. shortages that mirrored what had previously only been seen under the U.S.S.R.

I recall being on a call in March 2020 with a group of CEOs and the topic of food security, food availability, and agricultural supply chains came up. But it wasn't in the way you might be thinking. It certainly wasn't in the way I expected.

Cumulatively, the group controls tens of billions of dollars in annual revenue. And although food security was an important focal point of our discussion, the focus wasn't on the issue writ large at the population level; the discussion was about what companies and apps these executives could use *personally* to get food that was unavailable, including cheese, meat, fish, and produce.

This was an eye-opening moment for me. And it was clear that the economy and society at large was on a razor's edge that many of us had never seen — or even thought imaginable — in the OECD, let alone in the United States, which is one of the largest exporters in the world of food and agricultural products.

Not long after that group call, I spoke with a friend who is an executive who also owns a working 200-acre dairy farm. He was so concerned about meat supplies that he procured a side of beef, which is half a cow. And he told me he and his family had planted two to three acres of crops and had purchased baby chicks to raise chickens.

The fact that someone in farming, who owned a 200-acre working farm, was concerned about food supply chains was even more shocking than that executives running billions of dollars couldn't get a hunk of cheddar cheese or toilet paper locally.

I addressed these risks and incorporated some of these topics in my book *The Future After COVID*, which was released on 1 April 2020. Since that release, issues and risks around food and ag have remained top of mind for many Americans.

It also has frequently been a topic of discussion. When I spoke to homebuilders about the impact of the COVID-19 pandemic, I recommended they prepare for people to have more at-home food inventories than they have at almost any time in the past.

The implication? Build more and bigger pantries.

When I gave a presentation to the American Psychological Association in early May 2020, the topic came up as well. The APA board was most understandably interested in the psychological issues and changes people might experience as a result of the COVID-19 pandemic, shutdown, and recession. My observation was that people might have unhealthy relationships with food for some time. This is what happened for many after the Great Depression, and we might see something more severe given the normal abundance most people can access at any time. Of course, I made this observation not just as an economist looking at economic data showing increased at-home food purchases and a rise in food prices; it was based on the fact that everyone I know — every executive, leader, educator, and professional — was worried about food and was overbuying food and extra freezers and refrigerators as a way to extend their amount of food inventory in case food shortages became more extensive.

My birthday fell at the start of the shutdown. I spent that night like many others — and like many people at the time — lying awake in bed in the middle of the night, switching between food delivery apps, waiting for food delivery times to open right after midnight. The competition and agitated food frenzy were like the Cabbage Patch Baby craze of the 1980s, except it was for actual cabbage.

And then there were the food and toilet paper memes. Humor is said to be misery as seen from the other side of the street. But during the COVID-19 shutdown, it felt like there was only one side of the street, even though for some it was a much more dire situation, with more people becoming unemployed in the United States during the COVID-19 pandemic, shutdown, and recession than during the Great Depression. Plus, the asymmetric spike and stickiness of unemployment rates in communities of color have been a hallmark of inequality and have exposed social fault lines at risk in the future.

Media stories of meat shortages and other limited supplies were widespread. And while some might decry these as overblown, in truth, our entire economy and society was probably a lot closer to the edge than we would like to admit to ourselves.

MI5 — the British Security Service — has long held the motto that society is "four meals away from anarchy."[1] During the COVID-19 pandemic and shutdown, we weren't too far off. Furthermore, now that these risks have been exposed, they could be exploited again in the future as global power competition and conflict threaten to become more contentious.

I want to tell you that I was motivated to write this book solely by my interest in agricultural markets. To a certain degree, that was certainly a key driver. In fact, a book titled *The Future of Agriculture* was in development for many months before the COVID-19 pandemic. With the impact on food supplies and everything that happened in the wake of the COVID-19 outbreak, it made sense to pivot the focus of this research.

Fortunately, I have a solid background in analyzing and forecasting agricultural markets. In fact, I've been forecasting ag prices, analyzing ag markets, and assessing food and agricultural commodity risks for about 15 years. This has included forecasting agricultural commodity prices at Prestige Economics, for which I have been ranked the #1 forecaster in the world of cotton, sugar, coffee, and soybean prices.

But my career in the food and ag world has also included supporting ag commodity due diligence projects when I worked in investment banking at Wachovia, as well as directing McKinsey agricultural commodity risk management projects across many countries, continents, and products. This has covered everything from ethanol and biofuels to grains, coffee, sugar, and even more exotic topics like fruit risk management. After I got pulled onto a strawberry commodity risk management project, I thought I had heard and seen it all. But after COVID-19, that was clearly not the case.

This is partially why the impact of COVID-19 will cast a decades-long shadow on food and agriculture, as it will on many other industries that people had largely taken for granted.

Crafting futurist research, analysis, and scenarios are a cornerstone of my mandate as the Chairman of The Futurist Institute where I help analysts, strategists, and leaders become futurists. This is the primary reason I wrote this book: To share a futurist examination of the trends and technological innovations that will impact food and agriculture in the coming decade and beyond.

In general, my expectations for food and agriculture after the COVID-19 pandemic will be about widespread market adoption of existing technologies, penetration of emerging technologies into new markets, and ramping up bleeding-edge technologies, like quantum computing.

Furthermore, the future of food and agriculture in all its digital and virtual forms will be increasingly about security, including sustainability.

In the chapters ahead, I will discuss some of the most important trends that will rapidly become ubiquitous as well as some that face future challenges. I will also place these trends and technologies in their current and historical context.

After all, *this time is never different*, which is why we need to start with a discussion of the past of agriculture.

CRITICAL TRENDS

CHAPTER 2

THE PAST OF AGRICULTURE

Food is one of the most essential human needs, and for most of our history, obtaining food to consume was a core occupation and preoccupation of humans. It still is for many people, but in the most economically advanced countries of the OECD, it no longer occupies a majority of the workers.

In fact, in countries like the United States with its massive supermarkets, food and agriculture have become things that most people take for granted — or at least took for granted before the onset of the COVID-19 pandemic forced people to re-evaluate how secure they should judge their food supplies.

So, how did we end up with an abundance of food, an assumption of food security, and a populace no longer as occupied with food production?

Much of this can be traced to technological advancements. In some cases, relatively simple changes like crop rotation greatly improved yields. Horse-drawn plows had significant impacts as well. But the greatest changes have occurred more recently.

In fact, as recently as the mid-1800s, the majority of Americans worked in agriculture, as you can see in Figure 2-1. But the onset of the industrial revolution led to a rise of capitalism, urbanization, and an overall decline in the agrarian way of life.

There was massive wealth creation as well as a broad-based deployment of mechanized economic activities that allowed for much higher levels of productivity — in factories as well as in fields. This was further accelerated by the use of petroleum fuels in combustion engine vehicles in agricultural industries. In short, fewer people were able to produce more food. This led to manufacturing and the service sectors becoming increasingly important parts of the labor force.

Figure 2-1: Percent of Labor Force by Industry[1]

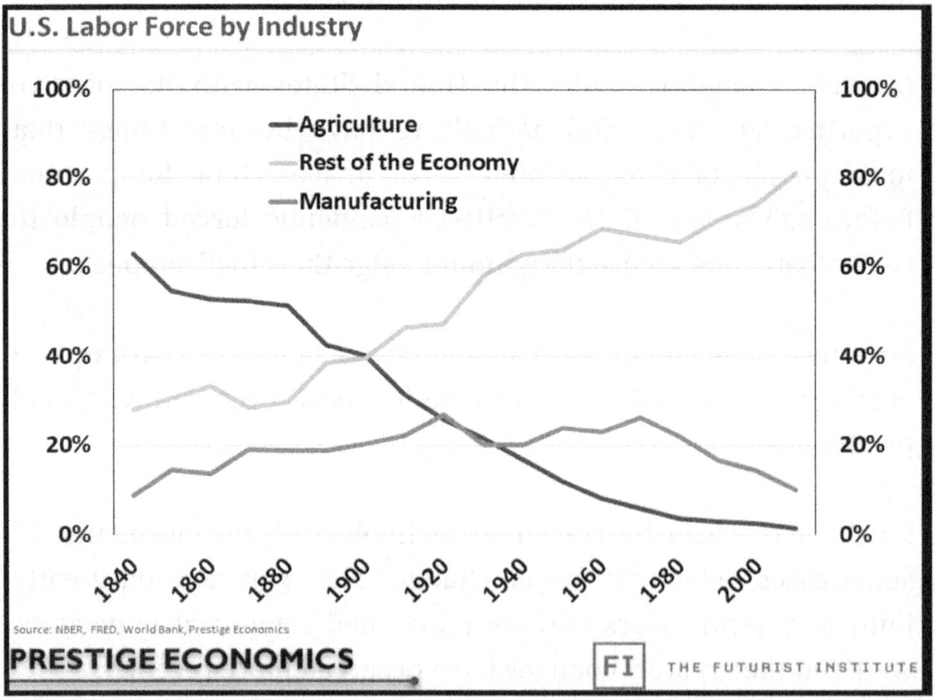

This became most significant at the turn of the last century and has continued to fuel the decline in agricultural workers as a percent of the workforce in the United States, the OECD, and globally. In 2019, a little more than 2.1 million workers were involved in agriculture in the United States, which is less than one percent of the population. You can see that this level has been relatively stable for two decades in Figure 2-2.[2]

Interestingly, U.S. agricultural workers represented just around 1.3 percent of the 2019 U.S. labor force.[3] Even though ag workers as a percent of the global workforce has been falling globally, the low U.S. level stands in stark contrast to the much higher percentage globally and in other countries, which can be seen in Figure 2-3.

Figure 2-2: U.S. Employment in Agriculture[4]

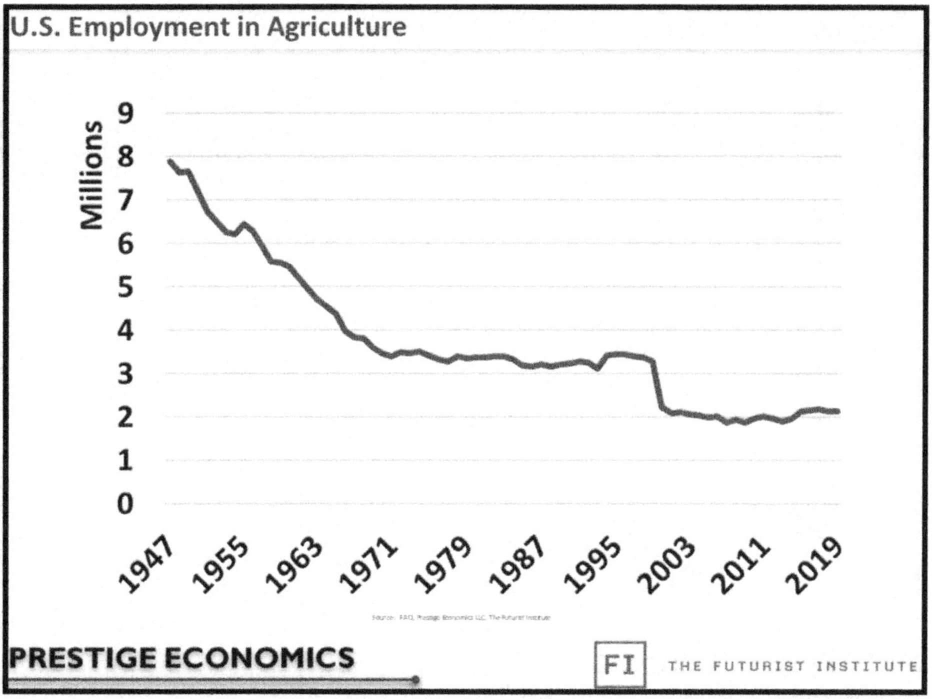

While some of the biggest declines in the percent of the U.S. workforce in agriculture happened in the first half of the twentieth century, there have been significant declines globally and across economies just since the end of the Cold War.

Looking at three important Asian economies offers a glimpse into this dynamic. In 1991, Chinese agricultural workers were around 60 percent of the workforce. In 2019, that figure is much lower at just over 25 percent. In Vietnam, the percent of ag workers fell in a similarly dramatic fashion from around 71 percent to around 38 percent between 1991 and 2019. But in the more advanced South Korean economy, the percentage of ag workers fell from around 15 percent in 1991 to 5 percent in 2019.[5]

Figure 2-3: Global Agriculture Workforce[6]

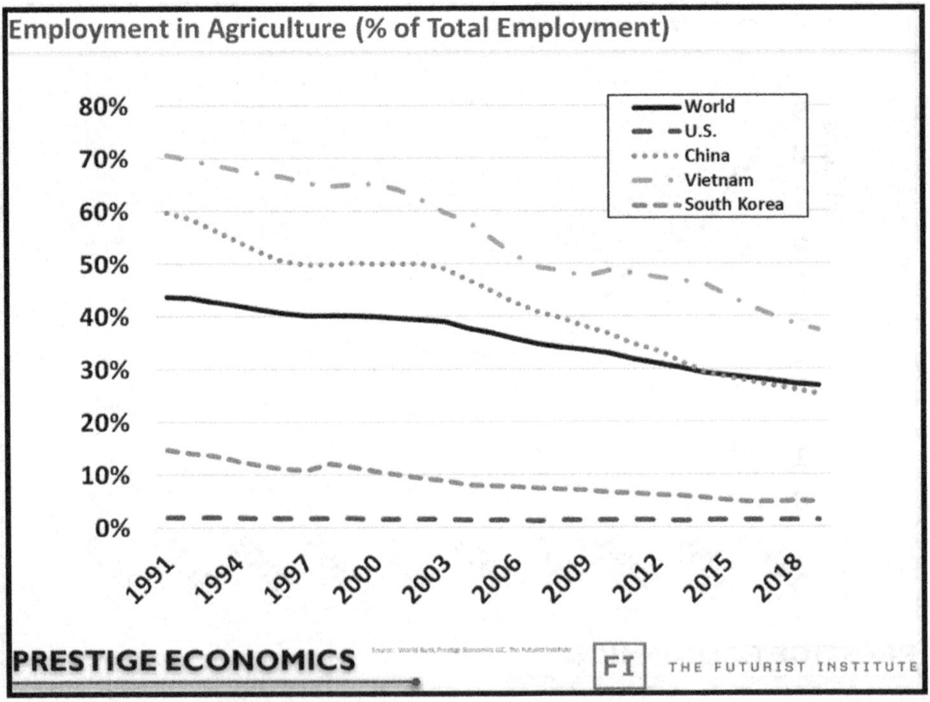

Meanwhile U.S. ag workers went from around 1.9 percent of the workforce to just 1.3 percent between 1991 and 2019, showing one of the smallest percentage point declines in the world.[7] This was largely due to the fact that the U.S. economy was already one of the most industrialized nations on the world in 1991. Logically, it would not have been reasonable to expect the same kinds of gains from emerging markets that would experience rapid industrialization, urbanization, and technological advancements as were seen in South Korea, China, and Vietnam in the past 30 years.

On a global basis, the percent of workers in agriculture fell from around 44 percent to 27 percent between 1991 and 2019. Looking ahead, this decline is likely to continue.

Even though it seems quite likely that more people in OECD economies may seek out careers in food and agriculture as a result of their COVID-19 pandemic experience, there are still productivity gains to be had in developing and emerging market agricultural industries. Furthermore, this doesn't even take into account the potential use of automated vehicles and drones in fields, lab-grown food, or highly-automated hydroponic facilities.

CHAPTER 3

THE PRESENT OF AGRICULTURE

If this book had come out before COVID-19, the present of agriculture would be a non-issue. I would have written that people take food security for granted and supplies are immensely abundant. I could have added that global trade and efficient supply chains provide greater access to goods across the planet than ever before.

But this isn't the pre-COVID world. And while some of the points about global markets are true — for now — regarding access and supply chains, the present is now impacted by COVID-19 supply chain bullwhip effects, inefficiencies, and pandemic labor instability. The biggest risks to global food and agriculture in the present aren't the food supplies themselves; the biggest risks are the people in the system.

Workers unable to process meat or deliver goods resulted in massive waste — as well as shortages. Now, spikes in demand driven by some panic buying during the worst pandemic in a century could engender oversupplied goods, causing more food and ag goods to spoil.

Those are the urgent and important dynamics of the moment. And they will likely continue to cast critical shadows into the future. But there are other hallmarks of the present state of agriculture and food beyond panic.

In addition to ag labor market dynamics, where an falling percentage of the workforce has provided the labor for most of the world's agricultural products, land use has also seen significant efficiency gains.

This has been especially true in OECD countries, although ag land use has risen in non-OECD countries. You can see this dynamic in Figures 3-1 and 3-2. Interestingly, in all four of these countries, the percent of the workforce dedicated to ag has fallen.

Figure 3-1: Land Use for Ag in the United States and China[1]

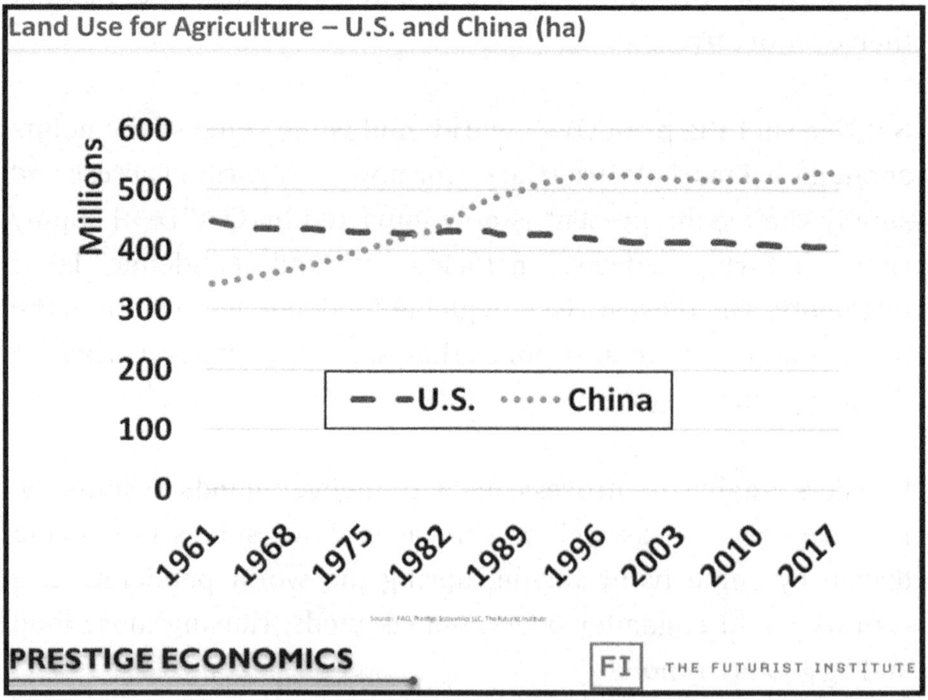

In Figure 3-1. you can see how U.S. land use has fallen in recent years, while Chinese use of agricultural land has actually risen.

In Figure 3-2, you can see how South Korean ag land use has fallen as Vietnamese land use has risen. Despite the massive rise in land use in Vietnam since the end of the Cold War, for example, the percent of the workforce dedicated to ag had fallen massively in that same time period.

These dynamics of productivity are remarkable. But even more remarkable is that the total acreage of ag land has fallen in recent decades, even as crop use for land has risen. This is just another indication of efficiency dynamics.

Figure 3-2: Land Use for Ag in Vietnam and South Korea[2]

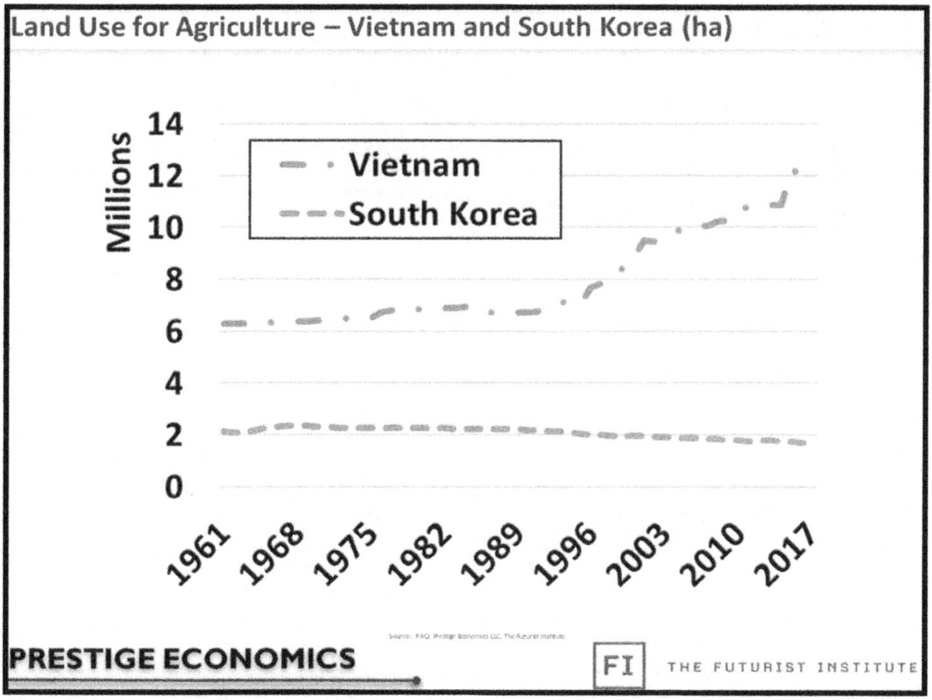

These data don't yet take into consideration the potential impacts of greater automation, sensors and IoT data, drones, or robots. But the efficiency trend of the past that has been continuing seems poised to extend beyond the present.

After COVID-19 has passed its peak pandemic period, the reality of food and agricultural abundance are likely to be visible. Even now, the fact that most societies were still able to get the calories they needed during the pandemic speaks volumes. But the supply chain and societal security risks that were exposed by COVID-19 are poised to foster an even-greater adoption of machines and automated solutions. In other words, the future of ag will likely be one that removes the risks presented by people — albeit one that will introduce new risks as well.

Figure 3-3: Global Land Use for Total Ag versus Crops[3]

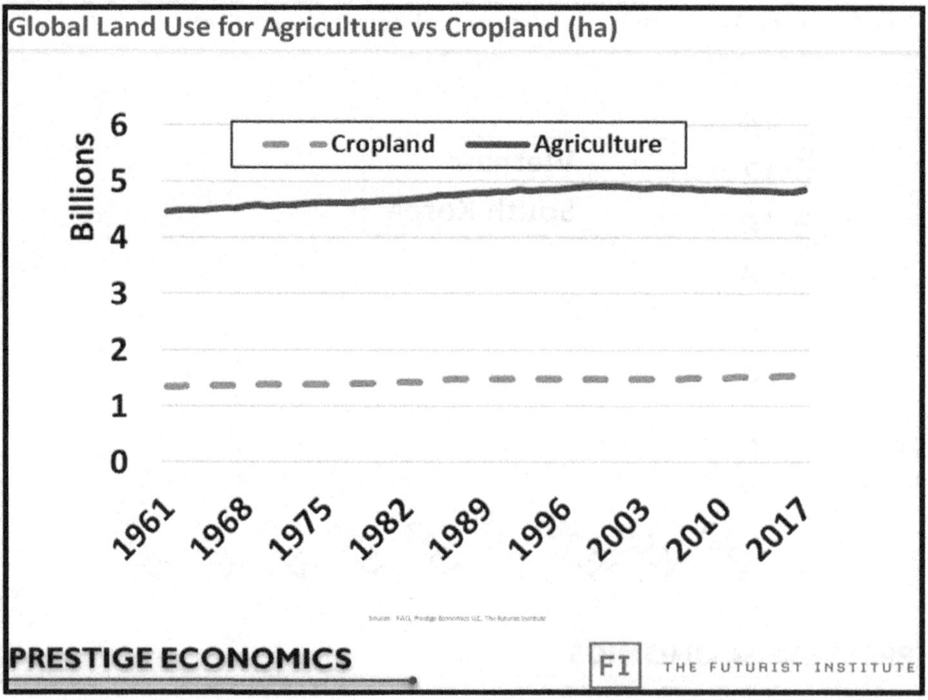

CHAPTER 4

THE FUTURE OF AGRICULTURE

This time is never different.

It's a saying often invoked by economists and futurists. It applies to financial markets, human behavior, and patterns of historical development and advancements in agriculture as well.

The past and the present of agriculture have leveraged technology to bring tremendous advancements in societal wellbeing, health, and wealth. Plus, workers in the most advanced economies have massively increased their real productivity in recent decades, but especially in the last decade of the twentieth century.

Looking ahead, a continued rise in global population is likely to drive a hunt for global calories and protein demand. This will stress the planet and sustainability issues will become more critical. But new and emerging technologies already exist that could help address some of the issues.

But before we consider the potential use of quantum computers that operate based on quantum physics ideas that Einstein referred to as "spooky," there is a chance to improve the efficiency of the cropland and acreage currently in use.

The trend of rising crop yields globally can be seen in Figure 4-1. Global yields have risen sharply in the past 60 years across economies, including the United States, South Korea, Vietnam, China, and globally. The fact indicated by the delta between these levels of yield by country GDP per capita is that there are likely further efficiency gains to be had with greater access to capital equipment and technologies already broadly in use in the most advanced economies. It will also likely require further ag industry consolidation of smaller farms in Asia and Africa.

Figure 4-1: Crop Yields[1]

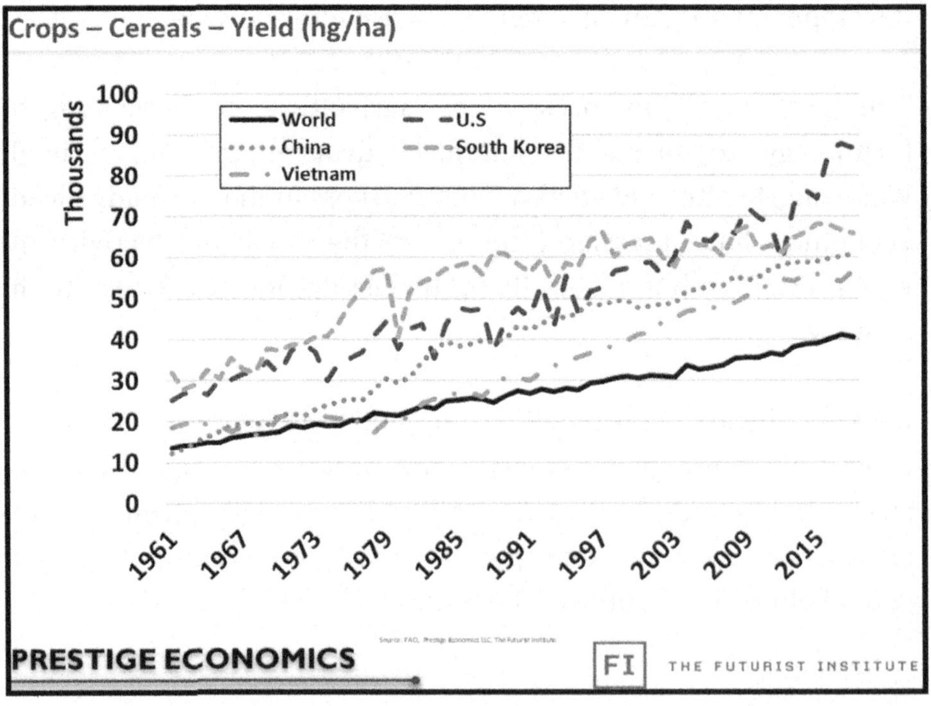

Of course, the crop yield of a country is significantly impacted by the weather and environmental conditions of that country. However, the differential between the yields in 4-1 have been consistent over time. And the multi-decade rise in yields across economies hints at the potential for countries to capture further efficiencies with existing technology.

During the time periods shown in Figure 4-1, the percent of the labor force dedicated to ag fell in all the economies shown. This, along with persistent yield gaps to the United States, which has a much smaller percentage of the labor force in ag, indicates that there is still a potential to substitute already existing technology and equipment for labor across economies and globally.

But what comes next?

This is the focus of the chapters ahead. The use of data, sensors, and more automation will likely allow for yields and agricultural productivity to rise across economies. Meanwhile, some technology, like quantum computing, could provide a step-change in optimization and blockchain may offer greater food surety and security.

THE FUTURE OF CALORIES

CHAPTER 5

THE HUNT FOR CALORIES

The biggest issue for the future of agriculture will be meeting the needs of a rising global population.

I'm not trying to be Malthusian here. I just want to be sure not to bury the lead, as journalists like to say. The big idea here is that agricultural demands in the decade ahead and beyond will stem from a rising global populace.

In fact, based on World Bank forecasts, there are likely to be more than 2 billion more people living on the planet in the year 2050 than there are now. You can see this dynamic in Figure 5-1. Furthermore, almost half of the growth in population is likely to be in the least developed economies in Africa and Asia.

But the rise in global population isn't the only additional source of food and agricultural demands in the decades ahead. After all, the rise in population is likely to be accompanied by a continued rise in global wealth; most importantly by a rise in an emerging middle class in Asia.

In Figure 5-2, you can see forecasts for global per capita GDP levels from OPEC. While OPEC's greatest concerns are about what global population growth and rising wealth mean for energy demand, we can extrapolate for the demand of agricultural markets as well. With massive rises expected in Chinese and Indian real GDP per capita, the demand for calories will likely rise as well.

More affluent economies have consistently demanded higher calorie diets. That dynamic is likely to persist into the future as well. Again, my point here isn't to argue that the world can't deliver on these needs. But it may require that crop yields reach their most optimal levels.

Figure 5-1: World Bank Global Population Forecasts[1]

Future Population by Region (Millions)

Country Name	2018	2020	2030	2040	2050	2050-2018
Europe & Central Asia	918	922	929	927	920	2
Least Development Countries	1,026	1,074	1,334	1,619	1,917	891
OECD Members	1,307	1,319	1,367	1,397	1,413	106
World	7,611	7,770	8,516	9,172	9,734	2,123

Source: Worldbank

PRESTIGE ECONOMICS

FI THE FUTURIST INSTITUTE

It will also likely require the deployment of technologies that are not currently broadly available to make sure the juice is quite literally worth the squeeze — or at least the water and fertilizer are worth the use.

Furthermore, it could require a slew of other technologies be deployed to ensure meat, produce, and other foods are available for a hungry Earth.

Figure 5-2: Real GDP per Capita Forecasts[2]

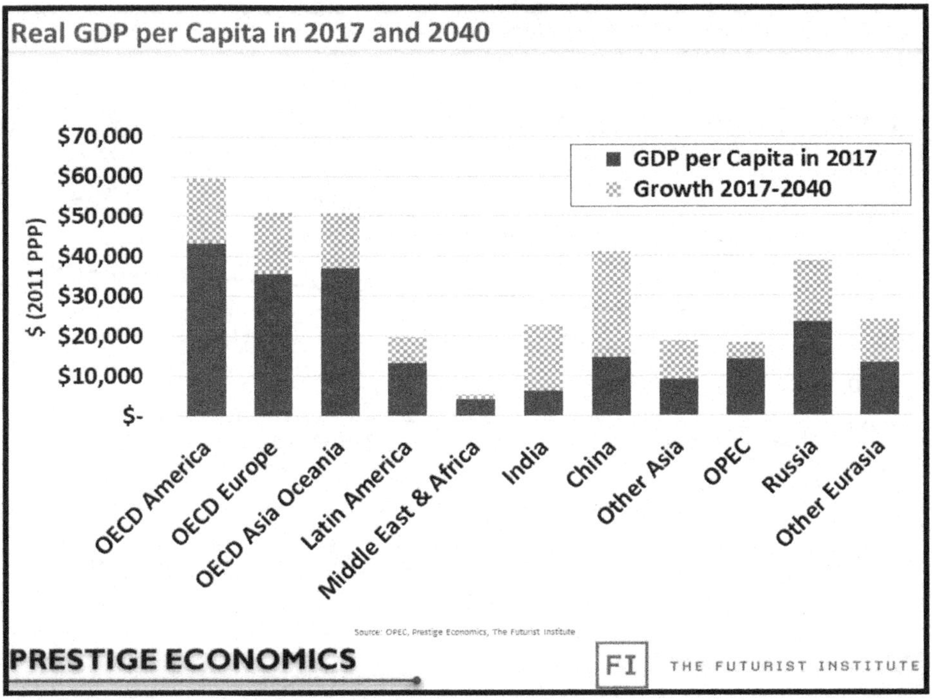

In Figure 5-3, you can see how daily diet calorie consumption varies across economies. And you can see that, on trend, calorie consumption levels have been on the rise.

Of the economies we have considered in a number of our graphs, the United States has significantly higher levels of per capita daily calorie consumption than South Korea, China, Vietnam, or the global average. Interestingly, the U.S. and Korea have the highest levels of per capita GDP shown, and the calorie consumption levels roughly fall in line globally with GDP output.

There should be no surprise here. The more people earn, the more daily calories they consume.

Figure 5-3: Select Country Per Capita Daily Calories[3]

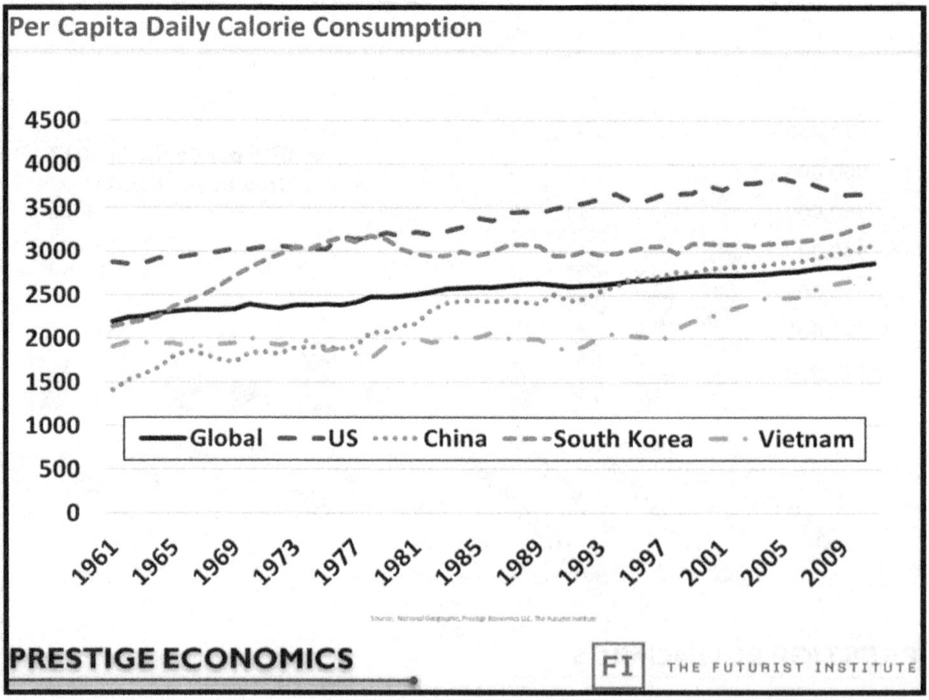

The trend rise in calorie consumption across all economies by economy type is something that has occurred globally for the world levels of consumption, as well as levels of consumption for industrialized, developing, and sub-Saharan Africa countries. This dynamic is shown in Figure 5-4.

Industrialized countries have generally appeared to be close to their maxed-out daily per capita consumption, even though the industrialized figure is still below the level of calorie consumption in the United States. This means that if 3,500 daily calories is the relative ceiling for per capita consumption, it could be a level that all global citizens eventually reach. This is 960 daily calories more than are currently consumed in sub-Saharan Africa and 450 calories more than the average global citizen.

Figure 5-4: Daily Calorie Consumption by Country Types[4]

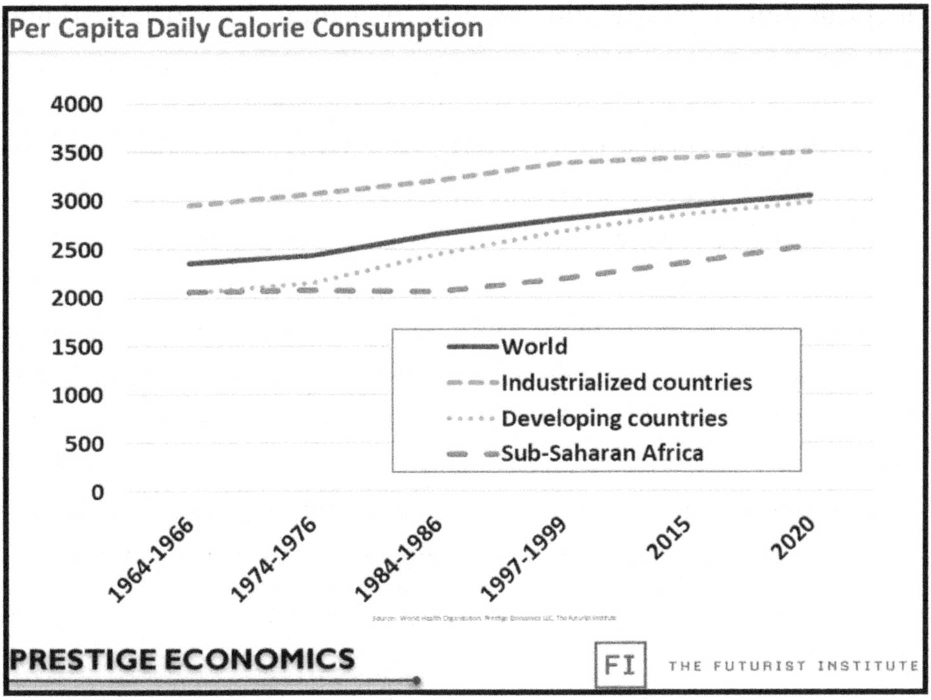

In other words, if the world average of calories is 450 calories shy of the industrialized level, we need to find an additional 450 calories per day for 7.8 billion people already on the planet for consumption to mirror average industrialized levels for every global citizen.

That's a lot of calories!

And it doesn't even take into consideration population growth or the fact that the additional 2 billion global citizens by 2050 might also optimize their food consumption at 3,500 calories per day.

In total, this means that the global economy would need to add around 10.4 trillion daily calories to meet the demands of over 9.7 billion people in 2050. To put that in perspective, the 2020 level of global calorie consumption is around 23.7 trillion daily calories.

In percentage terms, the increase of 10.4 trillion daily calories represents a need to increase current global daily calorie levels by around 44 percent by 2050 — in just the next 30 years.

Improved yields, increased land utilization, industry consolidation, and reduced food waste can help get supplies part of the way there. But the massive increases in demand through 2050 will require a lot more. This is why emerging and new technologies will be critical to push through a step change in calorie production to meet such a significant increase in demand.

CHAPTER 6

FUTURE OF PROTEIN

One of the greatest hallmarks of wealth in economies isn't just calorie consumption, it's also meat consumption.

As with total daily per capita calories consumed, daily per capita meat calories have been consumed at increasingly higher rates in recent decades. Even in the most advanced industrialized economies, meat consumption has risen on trend.

Of course, the most pronounced increases in protein consumption have come from some emerging and developing economies, where middle class wealth is emerging at an increasing pace. This consumption has been driven, in part by growing wealth and improved nutrition in absolute terms. But in other cases, it appears as if meat consumption has been prioritized in some economies — like China and Vietnam, where current daily meat calorie consumption exceeds even the daily meat calorie consumption in the home of the Big Mac.

In Figure 6-1, you can see how daily meat calorie consumption has varied across economies over time through 2011. You can see that, on trend, meat calorie consumption levels have been on the rise — with more pronounced increases for some emerging markets.

The United States is among the highest. Much like the calculation for global calories, the potential future needs for global meat could be staggering. In 2011, per capita daily meat calorie consumption in the United States was 469 calories. Meanwhile, the global level was only 272 calories. If we look ahead to 2050, it appears likely that global meat calories could double if consumption rises to the levels seen in the United States and other advanced economies.[1]

Figure 6-1: Select Country Per Capita Daily Meat Calories[2]

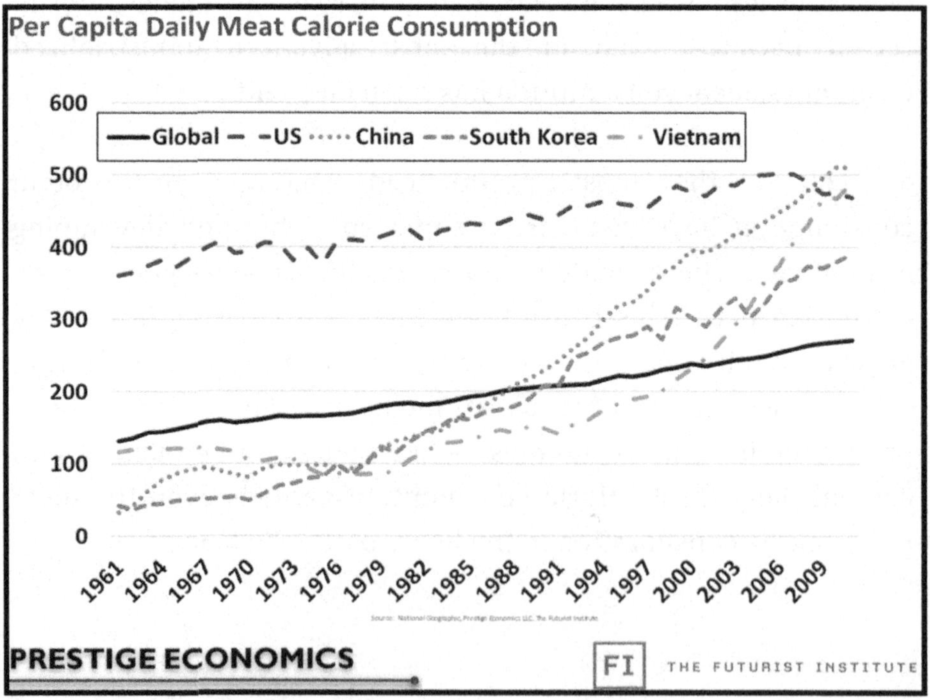

The emergence and rise of the middle class in Asia is a key factor behind the massive rise in global meat production in Asia in the past 40 years. Population growth is also likely a significant contributing factor.

In Figure 6-2, you can see that Africa is a very small portion of global meat production. The same was true of Asia 40 or 50 years ago.

Looking ahead, we expect that global meat needs will likely see the greatest increases from current levels in emerging markets that see increases in wealth, especially in Asia and Africa. This will be compounded by the rise in population in these regions.

Figure 6-2: Global Meat Production by Region[3]

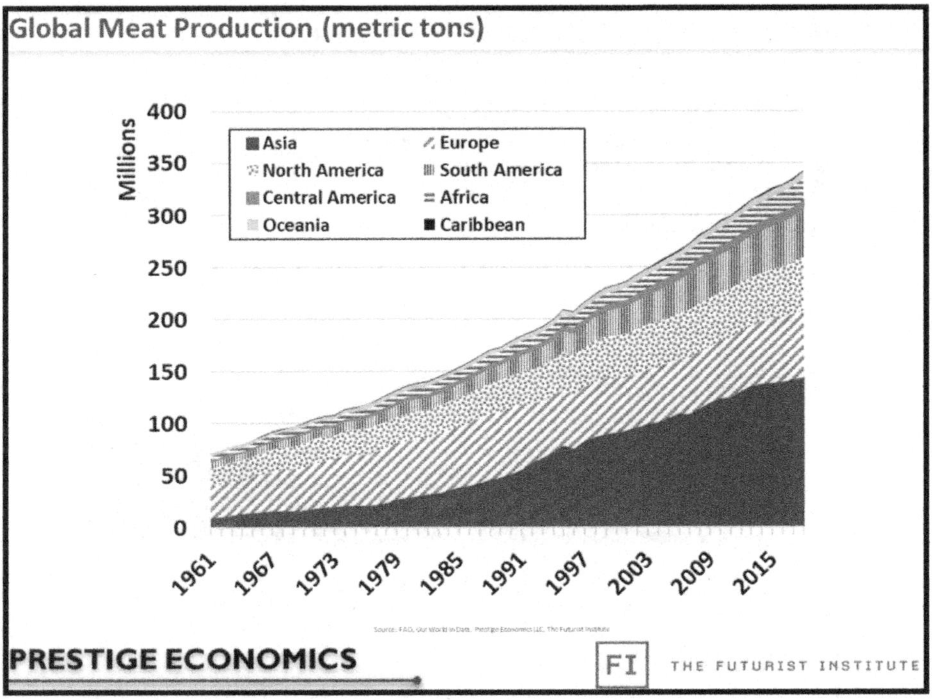

The biggest category of meat growth has been poultry. In Figure 6-3, the rise in poultry production coincides with the growth in meat production in Asia. Chickens are relatively inexpensive and they are relatively quick — between a month and a month and a half — to bring to market. As a stark contrast, beef can take between one and two years to come to market.

The implication from a supply chain standpoint is that if you want to convert grain to meat protein quickly, poultry is one of the fastest ways to do that. Of course, chickens are also really small. The impact of that dynamic is that to provide the pounds of meat necessary for the massive emerging global middle class, the number of animals slaughtered has gone parabolic.

Figure 6-3: Global Meat by Livestock[4]

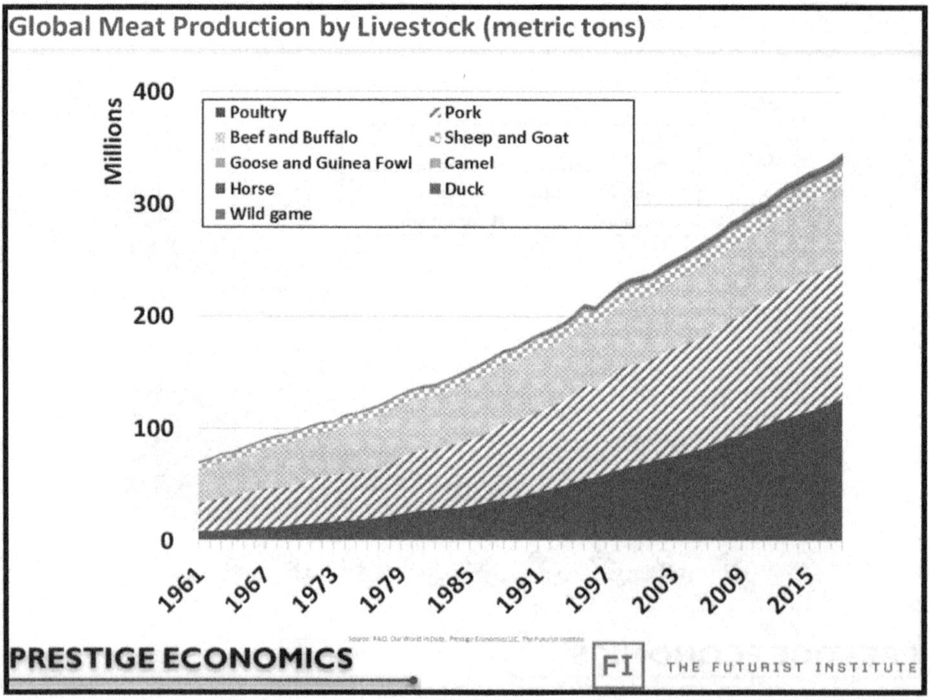

In 1961, the number of chickens slaughtered annually was around 6.6 billion. By 2018, it was over 10 times that at 68.8 billion. You can see this trend in Figure 6-4.

Looking at the future of meat, it seems that if we need to raise enough meat in the next 30 years to double the available meat calories on the planet, we might need to slaughter 120 billion chickens per year or more.

That's a lot of animals to slaughter, and the number could be even higher if fishing yields fall in coming decades due to overfishing, increased pollution in the oceans, and increased sea temperatures.

Figure 6-4: Animals Slaughtered for Meat[5]

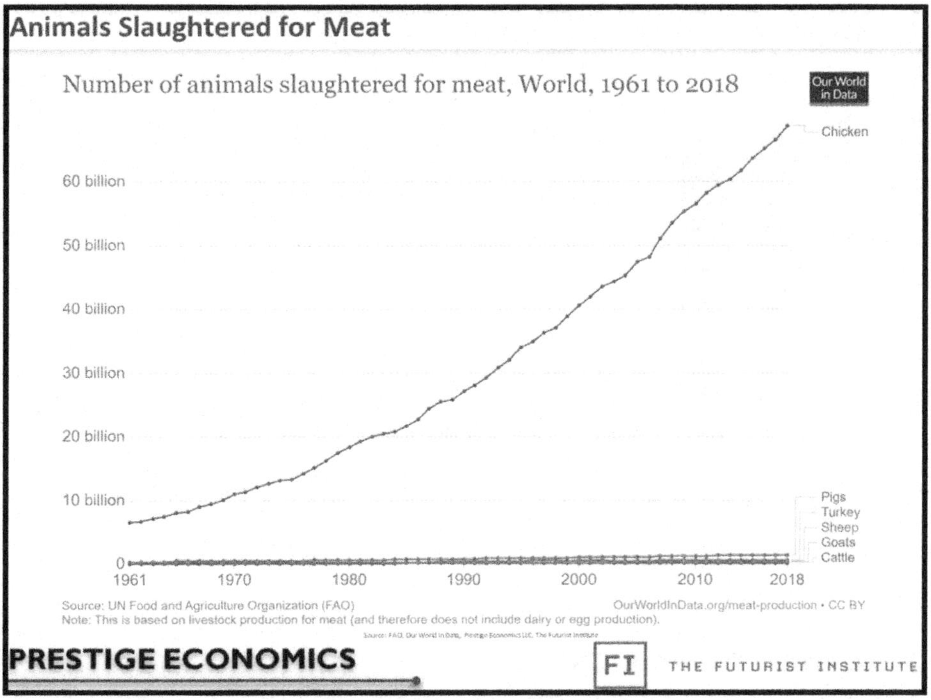

Beyond the sheer number of animals that would need to be slaughtered to generate the meat needed in the decades ahead, there are several other considerations including land use, energy efficiency, and greenhouse gasses.

Although poultry is far more efficient than other kinds of meats, as you can see in Figure 6-5, the sheer numbers of poultry required in the decades ahead is shockingly large — almost double current levels. The same is likely to be true of other meats that consume far greater amounts of land. In essence, as global wealth rises, demand for all kinds of meats is likely to rise.

This will pose issues from an acreage standpoint.

Figure 6-5: Land Use by Protein Type[6]

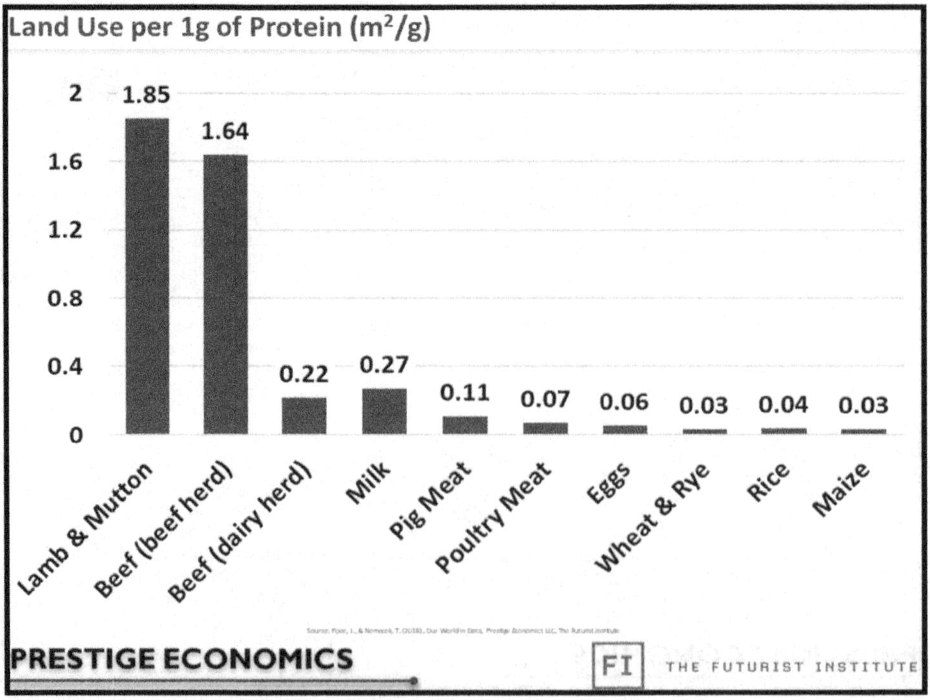

The energy efficiency of poultry would also continue to favor it as a source of protein — for both eggs and meat. Beef is likely to become increasingly rare (no pun intended), as acreage and land use demands are much higher for beef than other kinds of meat.

Plus, beef is the greatest producer of greenhouses gasses among proteins, as you can see in Figure 6-7. Poultry meat and eggs are again much further down the list. This should again make sense, since you get poultry meat in a little more than a month, but beef requires feeding a cow for up to two years.

That's just a lot more time to consume grain and water, produce greenhouse gasses, and take up space.

Figure 6-6: Energy Efficient Meat[7]

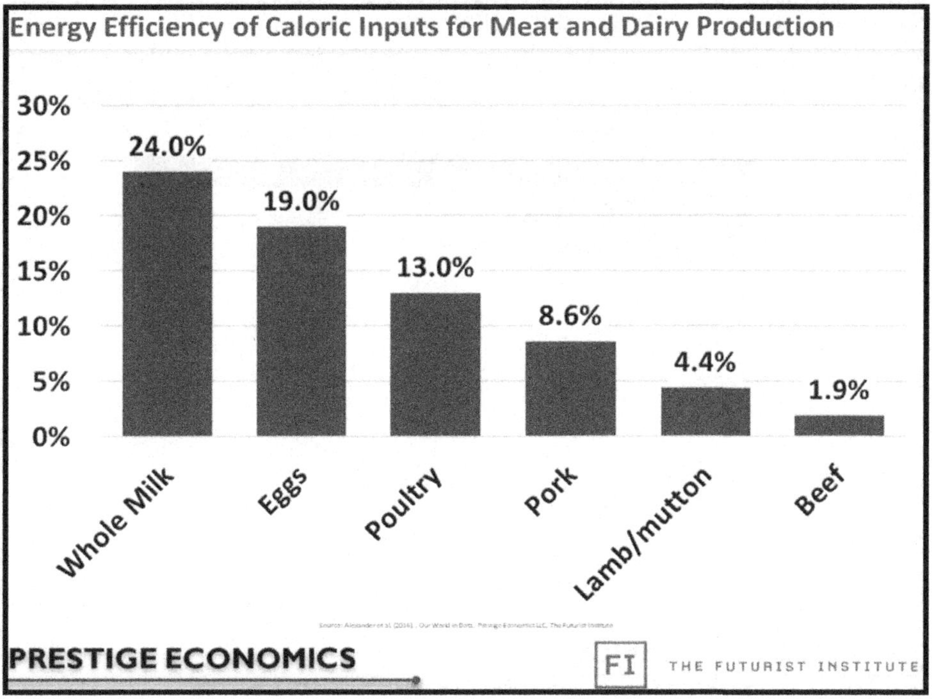

This isn't to say that beef or other meats will be outlawed globally, but if climate change issues continue to be critical for corporate initiatives, a war on beef could ensue that involves anything from a carbon tax on beef to corporate pushbacks against beef being served at meetings and events.

A pushback against beef would remove the greatest food culprit of greenhouse gasses from the mix. But it wouldn't solve the real problem, which is a rising global population and massively increasing global wealth. Those are realities, as is the fact that people need to eat and they need to consume protein. Furthermore, one of the accoutrements of middle class wealth is meat — especially beef in some places. This means demand could remain strong even with externalities priced in.

Figure 6-7: Greenhouse Gas by Protein Type[8]

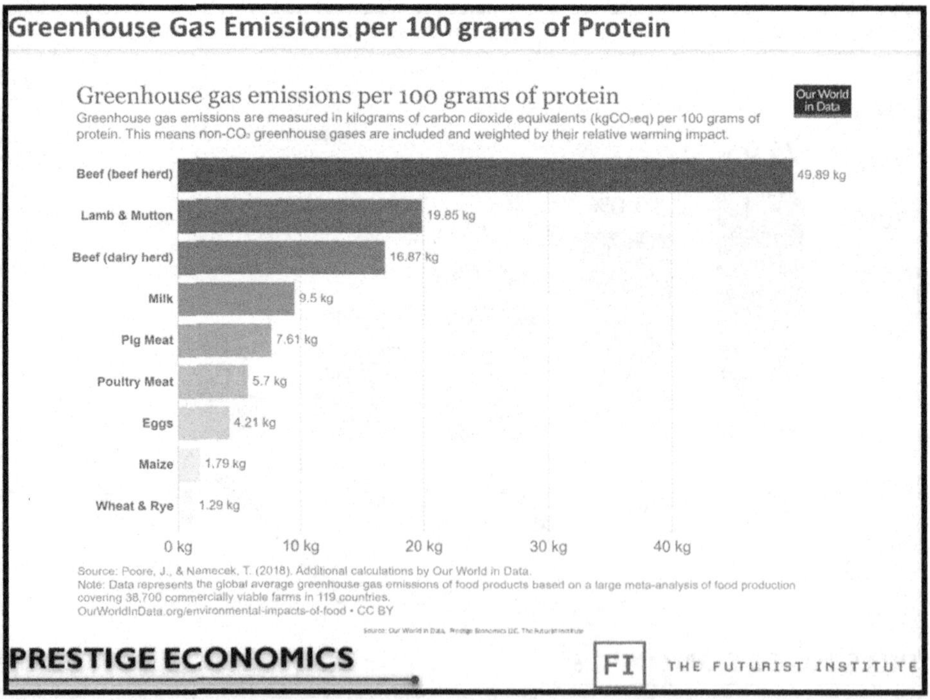

CHAPTER 7

GROWING MEAT

All of the meat data seem to point in the direction that in order to actually fulfill the demands of a growing global population, we will need to find ways to produce twice as much meat without consuming twice as much acreage or producing twice as many greenhouse gasses.

If meat production can be made more efficient and environmentally friendly, that would be a win. But if the solution to protein needs is to simply double the level of livestock raised on planet earth, the environmental impact on the planet could be devastating.

Fortunately, there are scientific plans in the works to grow meat, print meat, and integrate fish into hydroponic closed-cycle facilities to produce protein in a way that also minimizes water use and the environmental impact. This is going to be one of the most important technological steps forward for not only fulfilling the hunt for calories and satisfying the demand for meat, but also for sustainability.

Still, it sounds shocking to many to think that meat might be created in a lab or printed at home. I know the first time I heard the idea I though that I don't need a team of guys in lab coats in a multi-billion dollar lab growing my hamburger. After all, meat is relatively cheap and cattle can graze on grass that grows purely with sunlight and water.

But as the saying goes, "demographics is destiny." This means that the rising global population and rising global wealth will result in demands for greater quantities of meat, the likes of which may not be easily produced by traditional means. Plus, printed meat would not require the slaughter of 120 billion birds annually, let alone a few billion or so other livestock.

As for forecasts, there are a wide range of ideas about what the future of grown meat could look like. The market in 2025 is estimated to be $214 million,[1] but there are estimates out there that show over a third of all global meat will be "cultured" by 2040.[2]

In terms of the environmental impact, one study shows that lab grown meat produces 95 percent fewer greenhouse gasses, uses 98 percent less land, and consumes half as much energy as the farm-raised kind of meat.[3]

CHAPTER 8

HYDROPONICS AND THE FUTURE OF LAND

More likely than printing meat, which also just sounds quite literally distasteful, is the potential to harvest fish in closed hydroponic systems. These would involve plants in water tanks that also hold fish. The fish eat the roots of the plants and can be harvested for meat. Plus, the plants, which receive nutrients from fish waste, grow and yield fruits or vegetables that can also be harvested.

But that's not even the best part. The best part of hydroponics is that they don't just drastically reduce water use, they require much smaller amounts of space and could prove to be much more environmentally friendly than other options. It is not difficult to imagine giant multistory warehouses full of these self-contained systems that generate food.

This would be a much more efficient use of land and it could also reduce or mitigate the proliferation of food deserts in high-density urban areas by providing access to fresh food in locations without specific soil attributes.

Hydroponics and lab-grown meat may not be panaceas for addressing long-term trend increases in meat and protein demand, but looking ahead, the continued trend of increased protein production cannot likely meet the global demands of the next three decades without significant costs to the global system.

This will necessitate a push to alternative ways to source protein or else it will result in price inflation that becomes so unreasonable that people choose to become vegetarians and consume legumes and grains for their protein content.

The prospects for hydroponics are much greater than for lab grown meat, although hydroponics currently includes primarily non-meat products. According to one source, the global hydroponics market is estimated to be worth $16.6 billion by 2025.[1]

That would be eight times larger than the lab-grown meat market.

DATA TECHNOLOGIES

CHAPTER 9

E-COMMERCE AFTER COVID

COVID-19 exposed the world to e-commerce.

Sure, it has been around for a decade or so, but it was only during the COVID-19 pandemic and economic shutdown that people really learned what e-commerce was and what it could do. The intense lack of access to basic goods — and the inability to leave one's own home — resulted in people figuring out food delivery services and the like.

In many ways, this was a key part of the digital transformation for many individuals. The notion of going from a life of commuting and errands to one of both remote work and remote life was a potential that was already out there for many. My wife and I have been using online apps to order groceries for some time. But COVID-19 was the tipping point, where the rest of society figured out how to get groceries delivered and fully embraced ordering everything online — from musical instruments to workout equipment and clothes. But of all the things ordered online, food was most critical.

The trend of e-commerce retailer growth has been accelerating gradually in recent years. In Figure 9-1, you can see how this trend has been growing steadily in the past two decades. But this data also reflects that in the second quarter of 2020, the figure spiked sharply. During the COVID-19 pandemic and shutdown, e-commerce retail sales rose sharply to $211.5 billion in Q2 2020 from a much lower $160.4 billion in Q1 2020.

To highlight just how big a change this is, at the start of the last recession in Q4 2007, only around $36 billion of retail sales was provided by retailers primarily in e-commerce. The difference between Q1 and Q2 2020 was $51.1 billion in additional e-commerce retailer sales. In other words, this could not have happened in the last cycle.

Figure 9-1: E-commerce Retailer Sales[1]

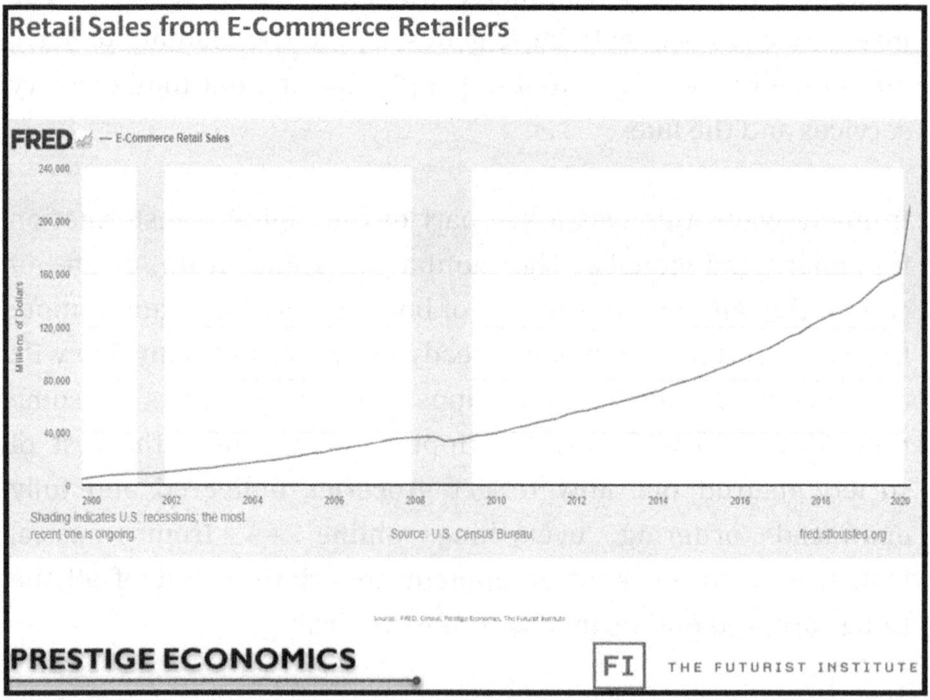

If we had the COVID-19 pandemic and shutdown during the mid-to-late 2000s, the societal disruption would have been much more significant, and the U.S. supply chain and e-commerce retailers would not have been able to step up and meet the needs of consumers the way they did in 2020.

E-commerce also consumed a much larger share of total retail sales in Q2 2020. In Figure 9-2, you can see how the percentage which had been steadily increasing shot up to 16.1 percent of total retail sales from just 11.8 percent in Q1 2020. These increases were significant, and the gains seen in the second quarter may not be easily reversed. For most people, goods they buy only bring value in their utility, and the convenience of e-commerce is not something they will give up easily.

Figure 9-2: E-Commerce Retailer Sales as Percent of Retail[2]

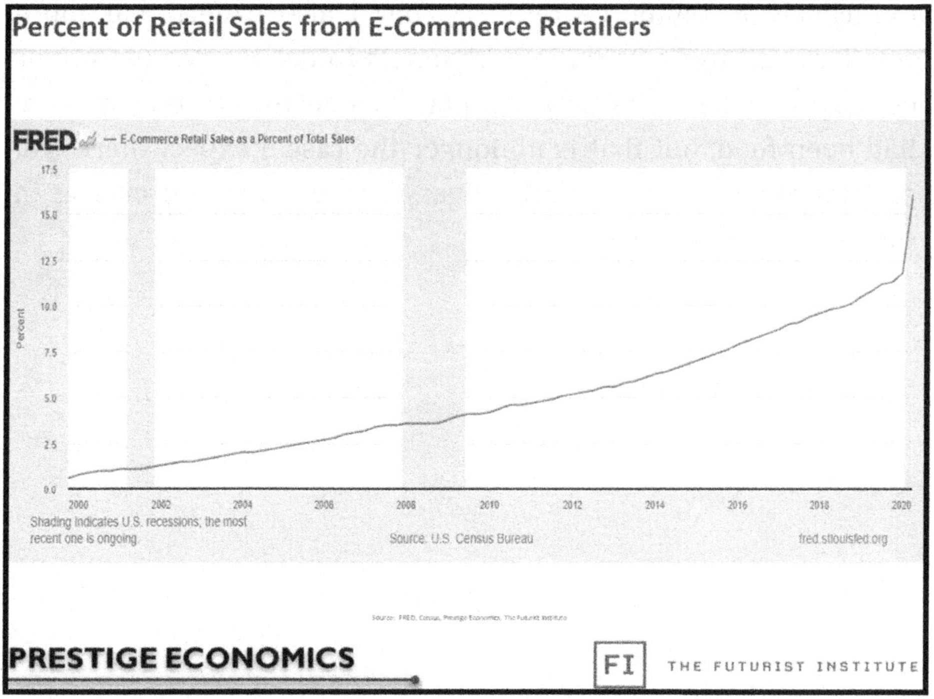

This means that people who value their time and convenience are likely to continue to consume from predominantly e-commerce retailers.

And there is one further caveat to the data here.

These e-commerce figures don't even reflect how much online e-commerce was performed by non-e-commerce retailers. They just reflect sales at companies that predominantly identify as e-commerce retailers. In other words, the impact of e-commerce is likely much larger than these data series even reflect. After all supermarkets and other stores that primarily operate as brick and mortar retailers don't show up in this data.

I expect more e-commerce to be done with retailers that are exclusively e-commerce retailers, but I also expect that more retailers will use e-commerce channels to deliver goods to their customers. One of the most impervious sectors to e-commerce had been food, but that is no longer the case. I expect there will be a lot of e-commerce food delivery as a percent of groceries in the future after COVID.

CHAPTER 10

USING DATA TO UNLOCK VALUE

The culture of data may finally have been ushered in by the COVID-19 pandemic and economic shutdown. Senior leaders who may have been averse to embracing technology have been forced to cross a Rubicon of technology. And one of the key artifices of the new world will be data. This will be true in food and agricultural businesses as in many industries globally.

Back in October 2018, I attended a speech where a Google executive noted that the volume of data collected between 2016 and 2018 was greater than the amount of data created and collected in all previous human history.[1]

Businesses live and breathe data. They are expecting to find their next customers, cut costs, choose their next initiatives, and optimize their activities by analyzing data. But the risk of computational data analysis paralysis is real if the data sets are too unwieldy for the computational powers at our disposal. Now, after being forced to embrace technology in an even greater way than before the COVID-19 pandemic, it seems likely that data collection, aggregation, and analysis will increase significantly.

The value of having data isn't in the data itself, but in the implications that can be derived from the data you collect and analyze. This all requires significant time and effort — and it also requires hardware to process the data.

Part of the reason for the massive shift and increase in data volumes has been a massive decline in the cost of data processing and data storage hardware. But that may be coming to end — especially for processing power costs. This is because computational powers are facing some potential limitations. In fact, many people working in technology will speak about a big near-term risk, and that's the limit of something called Moore's Law, which is named after Intel founder Gordon Moore. It is the notion that computational powers can be doubled while the costs are halved.[2]

This has made increasing computer processing power technology both more powerful and cheaper. A graphical depiction of the development of computer processing power along the lines of Moore's Law is depicted in Figure 10-1. But Moore's Law is breaking down.[3] The computational increases with lower costs just aren't happening anymore. In truth, the only solution now is more processors versus better processors. And this is a really big problem as the amount of data being created accelerates.

After all, if you want to be able to analyze data you collect, you need processors that can handle the amount of data you are collecting. This is going to become more of a challenge as the amount of data continues to rise parabolically, especially in an *Internet of Things* — or IoT — world.

Currently, we have an Internet of computers, tablets, and cell phones. Those are the primary devices that have Internet connectivity and through which data can be collected and tasks enacted.

But the drop in sensor costs and a desire for increased consumer convenience and business insights are going to usher in the era of the Internet of Things, when many things are connected to the Internet, not just your computer, tablet, or phone. These could include your car, refrigerator, and office supply closet. Looking out at the years and decades ahead, equipment, livestock, and physical plant assets used in agriculture will also be increasingly providing data into a broader system.

Figure 10-1: Moore's Law[4]

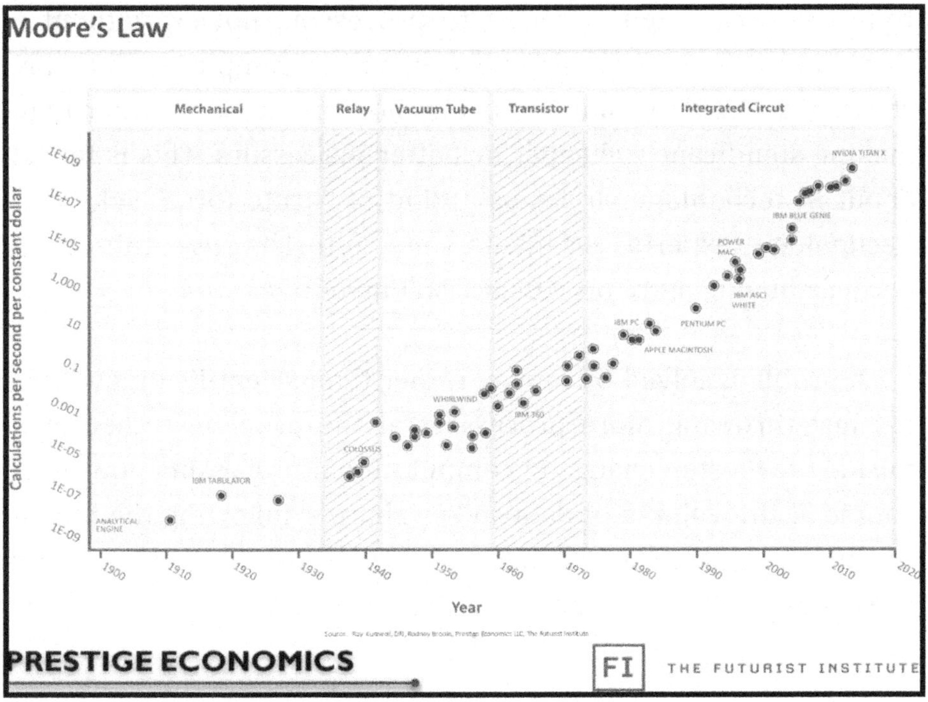

Each of these devices will interact, take actions, and generate data. All kinds of equipment and facilities will feed data about agricultural commodity prices, crops, weather, livestock, feed, the soil, equipment utilization and maintenance, and more into massive operational and computational systems.

When all these agricultural equipment and other devices are connected to the Internet, the amount of additional, minable, analyzable data will be mind blowing.

The speed at which this happens is also poised to accelerate as 5G technology proliferates in the years ahead. Current levels of existing data collection and analysis will likely pale in comparison to future IoT data.

Unfortunately, when it comes to processing power, without a step change from technology like quantum computing, the only strategy available would be buying more processors rather than making significantly cheaper or better processors. This is talked about in technology circles as a kind of "brute force" solution.[5] Technology experts, scientists, and futurists use this term because buying more processors isn't inventive.

It's a problem solved by money rather than scientific innovation. It's just throwing more processors at the problem rather than pushing for a step change in computational processing power. Of course, if the volume of data spikes sharply following the forced COVID-19 digital transformation of countless industries, brute force may be the only way forward in the short term.

Data Challenges Are Different

IoT devices will offer greater access to data in the future than in the present. And hardware may offer the processing power to make analysis more — or less — expensive, depending on whether universal quantum computing gets to commercialization quickly.

But the situation for data analysis is different.

You can throw as much money at the hardware as you like, but that isn't the biggest challenge when it comes to data analysis. Throwing more money and processing at significantly higher volumes of data just isn't enough. It won't help make sure the analysis is good. It won't solve the actual biggest data challenges.

These are some of the reasons that quantum computing, which I discuss in Chapter 12, will become a critical technology as the sheer volume of data we are collecting rises.

In the agricultural world, the amount of available data is already mind blowing. After all, weather, agricultural commodity price data, and crop data already exist in abundance.

Data Assets

There is one other final point that needs to be considered about the relationship between data and businesses. Some companies currently have line items on their balance sheets for data. This means that the FFA (Future Farmers of America) may need to be talking more about data when they talk about the future of farming and agriculture.

The evaluation and consideration of data as a valuable corporate tool is likely to become as important for many agribusinesses as the equipment out in the fields or the land itself. Public companies in agriculture and food are likely to increasingly rely on data, and the inclusion of data as a balance sheet asset is likely to increase in the years ahead for this field as well as all kinds of organizations.

This means that data isn't just something that professions and fancy Wall Street types need to understand in the world of ag. Farmers will also need to have a better handle on data. It means that as a resource, more data will have a positive impact on company operations, strategic planning, and valuations.

The notion that data is a valuable company asset that may be worthy of going on a formal balance sheet has been around for some time. It is likely to prove even more relevant in the future, now that there is a much greater potential for companies to embrace a culture of data following the COVID-19 pandemic — and now that investors are more likely to prioritize the value of data as a critical element for generating future cash flow for all kinds of businesses, including in food and agriculture.

CHAPTER 11

BLOCKCHAIN

Another technology that is already being used in the food and agricultural space, and is likely to be used with increased frequency and ubiquity, is blockchain. But before we dig in on the food and agricultural implications, let's take a look at what blockchain technology is and the origins of its sometimes dubious reputation that should not hinder its future use in food and ag.

Despite its existence for over a decade, when people talk about the future of finance, blockchain is often at the top of the list. The interest in blockchain was fueled greatly by the cryptocurrency bubble in late 2017, which had traders asking, "When moon? When Lambo?"

This mantra can be explained as embodying two questions:
- ***When will the value of my cryptocurrency go to the moon?***
- ***When can I buy a Lamborghini with my modest investment?***

But blockchain is about a lot more than Lambos! And there is a lot of misunderstanding and confusion that surrounds the topic.

Blockchain is like a combustion engine. In the way that a combustion engine can be used in any number of different vehicles, blockchain technology can be as well. At its core, blockchain is, in effect, a kind of database with specialized permissions and data sharing that has significant value for corporations and society.

This could be especially valuable when it comes to food safety and security.

Like other trends and technologies in this book, blockchain has been around for some time. After all, the first official Bitcoin transaction that was implemented using blockchain technology occurred in January 2009. Plus, databases and recordkeeping have been around for millennia. In fact, some of the first recorded writings are based on transactional records. So it should be no surprise that technological innovations that advance transactional recordkeeping would be important — or that they would be continually improving.

After the moons and the Lambos showed up in 2017, companies parroted their interest in the subject. They generally began to look at blockchain in terms of practical deployment and physical use cases in 2018 and 2019. This means that for the future of food and agriculture, considering the deployment and use of blockchain as a technology for recordkeeping and data transparency — and cryptocurrencies for digital payment purposes — will be an evolving process in the decade ahead and beyond.

But it is also unlikely to be a straight line from now to all crypto. This is because cryptocurrency blockchain technology was originally designed to circumvent traditional financial systems and structures as well as AML/KYC financial regulations.

Best in a Distributed Network

There are a few caveats about a blockchain that help elucidate its usefulness and purpose. The first of these is that blockchain works best in a distributed network. In Figure 11-1, you can see what a distributed network looks like. It's when you have a number of different parties engaging in transactions and there is no central organization to the nature of these transactions. This is like what happens with regionalized businesses or what might happen with how members of a family might spend their income. You have independence of actions and no natural central repository for all records.

Figure 11-1: Distributed Network

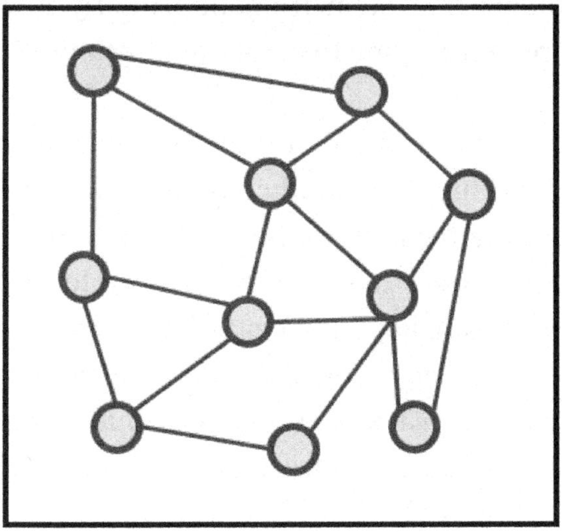

Of course, a set of records can be created and shared with a distributed network for any kind of transaction. But this has historically proved clunky. Anyone who's ever been cc'd on an email chain for a major project knows how difficult it can be to keep track of every update. Plus, sometimes the wrong files are updated, and this can lead to errors.

When I worked in management consulting at McKinsey between 2007 and 2009, this was referred to as a version control error. It was a major risk in almost any project or client engagement.

Sometimes people would update the wrong file, and then important documents were at risk of being incomplete or overwritten. And they would subsequently necessitate complicated reconciliation and review. This is also a problem in accounting, which is why for audits, tracking financial documents is critical. For auditors, ensuring a consistency of the data through proper sampling and testing is designed to reveal the kinds of problems that stem from improperly kept records — records that are more prone to errors in distributed networks.

Of course, fixing the risks of record keeping in a distributed network is something that cloud computing was designed to do. In cloud computing technology, people in different locations can edit documents that are updated in a shared location. But those documents are usually kept in a centralized location, and they do not always keep an easy-to-track record of who updated what and when. In some ways, this makes cloud computing like a centralized network, as in Figure 11-2, where all permissions and data are kept in one place.

It also means that the network can be vulnerable to a central point of failure. A central point of failure risk is one in which records are at risk because they are kept in one place — like Dropbox or Google Drive. After all, if Dropbox or Google Drive stop working due to a technology problem, it also means that you may not be able to access your documents. And your system of documentation and record keeping will fail through no fault of your own, at least temporarily.

Blockchain technology is different than cloud computing because it is explicitly designed to allow people within a distributed system to receive updated, permanent ledgers of completed transactions. It is designed to reduce central point of failure risk. When keeping track of food and ag products, knowing where things are and where they have been is critical for food safety.

Figure 11-2: Centralized Network

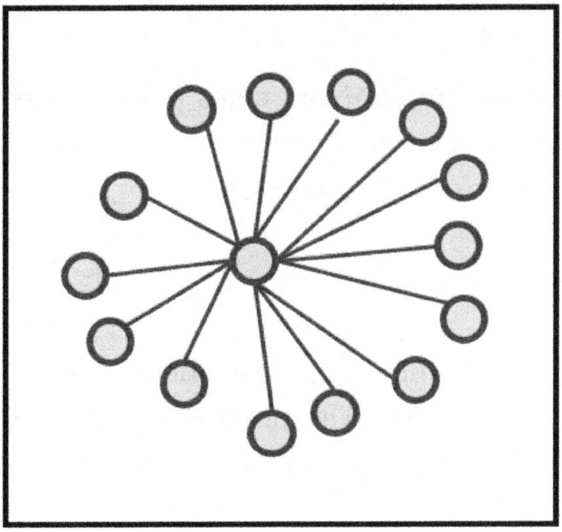

But safety isn't what blockchain's initial users were focused on.

After all, they were focused on the whole cryptocurrency piece — and those darn Lambos! But before we talk about crypto, let me just reiterate that blockchain is the technology engine that powers Bitcoin and other cryptocurrencies in a permanent distributed ledger.

But blockchain can power many things, including ledgers that track agricultural products and facilitate transactions and recordkeeping of those products.

Die Hard Crypto

Maybe it is because the name Satoshi Nakamoto is fake and it sounds Japanese. Maybe it's because people use crypto for faceless, nameless, anonymous transactions that it evokes another fake Japanese-sounding name for me: Nakatomi Plaza.

This isn't a real place either.

It's where the first *Die Hard* film takes place — where the villain of the movie, thief-cum-terrorist Hans Gruber, played by Alan Rickman, is trying to steal bearer bonds.

And what are bearer bonds?

Well, that's the really interesting part: Bearer bonds are a kind of bond that no longer exists. They are called bearer bonds because they are owned by the person who bears them.

If you have them in your possession, you own them. If you don't have them in your possession, you don't own them.

You can see a U.S. bearer bond in Figure 11-3.

While bearer bonds were first issued in the United States in the second half of the 19th century, they stopped being issued in 1982.[1] Today, they are all but extinct.

The reason? Anti-money laundering.

Figure 11-3: U.S. Government Bearer Bond[2]

Anti-money laundering, or AML, as most finance people know it, is the set of legal initiatives taken by governments and financial institutions to prevent and hinder terrorists, criminals, and other bad actors from being able to engage in actions by making it difficult to transfer, hide, and otherwise use money.

And how was it that these bonds were so convenient for money laundering? Well, since these bonds were owned by the person who held them (i.e., the bearer), and there was no trading account or electronic or paper record required to redeem them, they could be bought, sold, and traded in an untraceable manner.

This meant that bearer bonds could be used for money laundering and all kinds of other nefarious things. It's the reason the villain in the *Die Hard* movie wanted to get them, because they could not be traced, and they could be used for illegal activities.

Without any record keeping, bearer bonds could be easily used in illegal commerce without concern of detection because you wouldn't need to go into a Chase Bank branch or log into your Fidelity trading account to buy and sell them. You would, quite literally, take a suitcase of money and hand it to a fellow mobster or terrorist who would, in turn, give you the bearer bond.

So why do bearer bonds make me think of the interplay between Bitcoin and *Die Hard*? That's simple. Bitcoin and other cryptocurrencies can be used namelessly. Facelessly. Anonymously. They are digital bearer bonds.

Like the analog, physical bearer bonds before them, these digital bearer bonds are only owned by people who have the digital key. And they can be used for nefarious purposes. But while bearer bonds are all but extinct because of AML initiatives, regulation has seriously lagged behind crypto. Bitcoin and cryptocurrencies are ascending, and they have not been outlawed in most countries despite their ease of use to violate AML laws.

The irony of cryptocurrencies, of course, is that the idea of nongovernment currency started as a vision of Austrian economists and free market libertarians. Yet while these same libertarians may praise the freedom of operating a currency outside of government, if their Bitcoin gets stolen, the police and the FBI will surely be their first phone call as a recourse of action.

But it may all be for naught. After all, even though transactions on the ledger are permanent, they are difficult to track down because of the built-in anonymity of cryptocurrencies. And even if improving forensics could help trace ill-gotten crypto leveraging the records of a permanent ledger, stopping transactions as they happen would still be quite impossible.

So, as paragons of economic theory dream of the freedom crypto brings, the Hans Grubers of the world — including ISIS, anarchists, political subversives, organized criminals, and the rest of the digital underworld — rejoice in the ignorance and naiveté of those dreams.

Policymakers Know

Fortunately, policymakers and governments have figured out the potential problems with cryptocurrencies as an anonymous means for bad actors to circumvent regulated structures.

Christine Lagarde, the head of the International Monetary Fund, spoke on the topic of cryptocurrencies at a conference titled Terrorism Financing: The Other War Against Al-Qaida and Daesh.

At that event on 26 April 2018, Lagarde noted:

> **[FinTech] can be used to promote and fund terrorism through the anonymity of crypto-assets.** But leveraged effectively, fintech can also be a powerful tool to fight terrorism and its financing.[3]

The Bank of International Settlements, which is considered the central bank of central banks, noted in June 2018:

> They lack a legal entity or person that can be brought into the regulatory perimeter. **Cryptocurrencies live in their own digital, nationless realm and can largely function in isolation from existing institutional environments or other infrastructure.** Their legal domicile — to the extent they have one — might be offshore, or impossible to establish. As a result, they can be regulated only indirectly.[4]

The big takeaway is that cryptocurrencies exist outside of legal frameworks, and their anonymity facilitates terrorism funding.

In August 2020, the U.S. government seized millions of dollars in cryptocurrencies from over 300 accounts from al-Qaeda and two other terrorist groups that were using social media to attract crypto funds.[5]

For now, the cryptocurrency story has yet to fully play out. There are still many questions that need to be answered for regulators to give approval. And there is no guarantee that their complete approval will ever come. Furthermore, any cryptocurrencies that attempt to circumvent or subvert current, dominant, global finance structures may find themselves on the wrong side of regulation for a long time to come. The murky nature of cryptocurrencies is one of the reasons Facebook's plan to create Libra as its own cryptocurrency was so hotly watched by lawmakers and regulators in 2019.[6]

Assessing the Future Potential of Blockchain

The potential impact of blockchain is bigger than most people can imagine but not nearly as great as some believe. Yet, while many people think blockchain is just crypto, that's only the tip of the iceberg. This was a key part of a talk I gave at SXSW in March 2017, and it was the cornerstone of a book reading I did for *The Promise of Blockchain* at SXSW in March 2019.

The impact on corporate supply chains — on logistics, transport, and freight — is going to be massive. And there are a number of other industries, like finance and agriculture, where a chain of custody of ownership and good record keeping is likely to provide a significant value proposition for blockchain.

After all, blockchain allows for a permanent distributed ledger, and if used in a private, commercial endeavor, it could provide instant transparency of origin, content, and custody, which is often required from a regulatory framework for conflict minerals, chemicals or chemical content, or trade. Plus, this kind of transparency can all have significant value from a public health and safety standpoint, as is the case with agricultural products and food safety — an area that has been at the forefront of blockchain use and adoption.

The potential for blockchain to add economic value in many different industries and corporate fields is massive. But it isn't equal across industries. Plus, in some cases the ROI to deploy blockchain may not be present. In general, it appears as if the biggest value proposition for blockchain is where there are supply chain risks or health and safety issues to mitigate.

Assets in Transition See Greatest Use Case

The value proposition in using blockchains for record keeping on long-lived assets is lower than for assets that move. This is true whether we are talking about financial assets that frequently move between investments, markets, and parties — or assets that physically move. This impacts The Futurist Institute's assessment for blockchain use potential in Figure 11-4.

In addition to movement factors driving the use case and value proposition for blockchain, there are limitations on blockchain's potential uses in some industries due to incumbent regulatory requirements and dominant legal frameworks — especially in the United States.

This is why healthcare, property, and government data uses for blockchain may prove more limited. It also explains why there is a high potential for blockchain use in agricultural products, transport and logistics, and financial services.

The Futurist Institute's Blockchain Assessment

As the Chairman of The Futurist Institute, I directed an analysis and study of corporate and commercial opportunities in blockchain across industries. The overview across industries is in Figure 11-4, and agricultural products are at the top of the high impact and high potential list. Let's look at the reasons why blockchains in ag and food are likely to be more impactful than blockchains in finance.

Figure 11-4: Assessment of Blockchain Potential

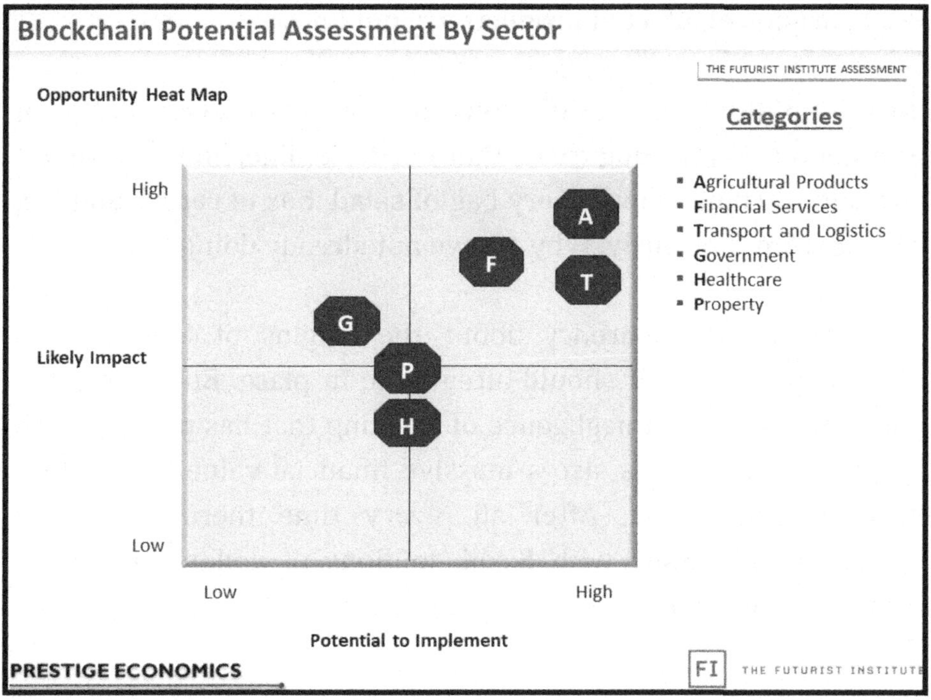

The Futurist Institute's Agriculture Assessment

Finance gets a lot of attention when it comes to blockchain use and potential. But one of the greatest industries with the potential for blockchain use is in the field of agriculture. As in the areas of finance and transport, the use cases and value proposition in agriculture are tied to transaction frequency.

Of course, blockchain isn't going to help you milk cows or grow corn faster. But it is likely to be important for tracking the origin of foodstuffs by knowing its custody and by having a transparent supply chain that facilitates trade and transactions.

The high public value — and economic value — of blockchains in agriculture is highly documented. Every time there is some kind of food contamination, it becomes a national crisis, because the exact origins of affected products are not known.

In my opinion, this is ridiculous. If I can track every bottle of shampoo or bag of dog treats that I order online, how are we not already tracking where every bag of salad, box of cereal, and egg comes from? Seriously. Why are we not already doing this?

The lacking transparency about the origins of foodstuffs is shockingly absent. It should already be in place. But even if we discount this abject negligence of tracking that has public health implications, there is also a massive financial value proposition from tracking food. After all, every time there is a food contamination issue with E.coli, millions of dollars of food is simply disposed of.

This is a waste on so many levels. And blockchain can help to ameliorate both the food safety risks and the risks of financial loss. This is reflected in The Futurist Institute's expectations for agriculture and blockchain in Figure 11-5.

The Futurist Institute's Transport and Logistics Assessment

In many ways the use cases for blockchain in food and agriculture are an extension of transport, logistics, and freight use cases for blockchain, which are shown in Figure 11-6. These use cases hold true for food and agricultural products as well, although food and ag have an additional safety component to their consumption that is more critical than if we are talking about blockchains tracking boxes of t-shirts or chairs.

Figure 11-5: Agriculture Blockchain Assessment

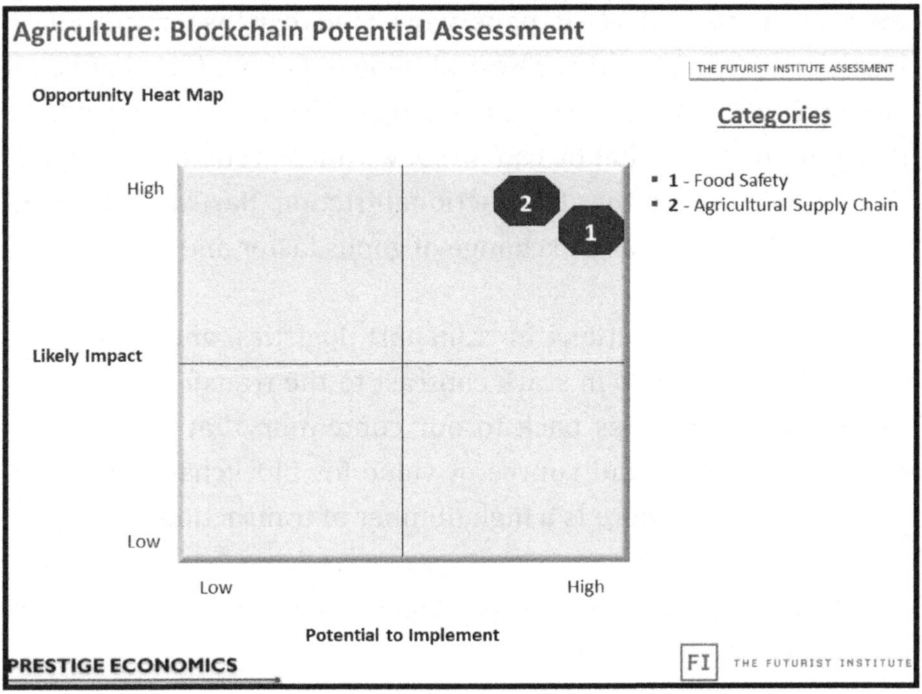

There are two big areas where we see blockchain use cases for transport and logistics that could also be helpful for agricultural supply chains: increasing the ease of physical transport and exchange of goods as well as regulatory compliance. Both of these depend on keeping a clean chain of custody and having easy-to-demonstrate records about the content and origination of goods.

This reflects a focus on corporate transactions as both the highest value and greatest importance from record keeping and commercial standpoints. In other words, using blockchains for these purposes is likely to bring significant value to large entities — and the economy overall.

Transport, logistics, and freight are industries with a high degree of what economists call *transactional friction*. This means that there is a great deal of paperwork and expense required to facilitate trade and the physical movement and exchange of goods. This is an ideal use case for blockchain, which provides a distributed record that be can be set within a predefined network to ease and reduce that transactional friction. Basically, it would make the transport and exchange of goods faster and cheaper.

Like finance, transactions in transport, logistics, and freight are frequent. This stands in stark contrast to the transfer of physical properties. And it ties back to our contention that the greatest potential use case and source of value for blockchain will be in industries where there is a high number of transactions.

Furthermore, like finance, some aspects of freight and transport are regulated.

This is especially true for some manufacturing restrictions on the use of some chemicals and conflict metals. Plus, counterfeiting is a significant problem for some high-value goods in the supply chain, like pharmaceuticals, certain manufactured goods like inverted delta parts, and military weaponry. And there are also high tech goods, like smartphones. Blockchain use could increase transparency in these supply chains, reducing potential fraud, increasing safety, and saving companies money.

From an oversight and regulatory standpoint, I expect a high potential for blockchain implementation in transaction monitoring including around food and agricultural products — especially if the U.S.-China Trade War that started in 2018 continues or becomes more severe.

Figure 11-6: Transport and Logistics Blockchain Assessment

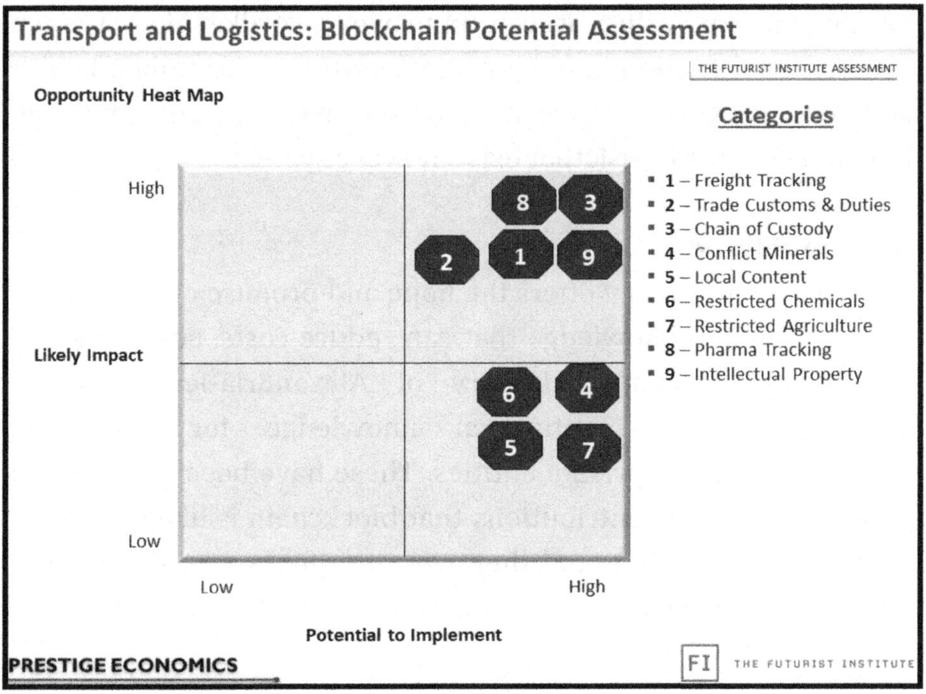

Trustless Trust is Too Far

People often invoke blockchain as a technology that fosters trustless trust, but that is simply going too far. Blockchain is not a 100% guarantee that the products involved are tamperproof or authentic.

There is still a need for people involved in the network to behave honestly and ethically. If someone in the system is a bad actor, there will still be adverse economic consequences — and in the case of food or agricultural products, there may be negative health outcomes. This means if we trust in blockchain absolutely, the consequences could be even greater than in our current system.

This does not mean blockchain should not be used. It should, because it has value, it can improve recordkeeping, reduce transactional friction, and contribute positively to public health and food surety and security. But we need to remember the human elements of risk that exist in every system.

Hope and Hype

Blockchain technology offers the hope and promise of distributed information and knowledge that can reduce costs, add economic value, and prevent a Library of Alexandria-level loss of information and institutional knowledge for corporate, governmental, and private entities. These have been some of the greatest potential contributions that blockchain is likely to make in the years ahead — and they were the focus of my book *The Promise of Blockchain*.

But blockchain is still also seen as hype. It is still seen as Lambos. After all, it has been tied to the still-dubious world of cryptocurrencies and the biggest investment bubble in the history of the world. Yet blockchain has the potential to significantly impact a number of industries in a positive way — especially agriculture and food production, supply chains, and consumption.

That is why I see so much promise for blockchain as an emerging disruptive technology that will impact food and agriculture in the decade ahead.

CHAPTER 12

QUANTUM COMPUTING

Quantum computing is poised to usher in a computational step change in the decade ahead — including in the fields of food and agriculture.

Essentially all computers in 2020 use binary computational programming of 1s and 0s. These 1s and 0s form *bits*, which stands for binary digit. But quantum computing is not just 1s and 0s, which are often seen as allegories for a bit that is either "on" or "off." In a quantum computing world, there are bits with other states of being between 1 and 0, called *qubits*, which can be either on, off, or both on and off at the same time.

As for the impact of quantum computing, adding additional states of being to the computational process may not sound like much, but the exponential impact for computing power is likely to be absolutely massive, which is why it could be important for the future of food and agriculture, supply chain, finance, blockchain, and other technologies.

Quantum computing could prove to be a critical technological step change that will materially impact business, science, communications, cybersecurity, and national security. But there are limitations, risks, and challenges ahead for quantum computing to achieve a true level of commercialization viability.

While some industries will benefit from specific new and emerging technologies, quantum computing has the potential to impact a wide range of industries in a general purpose fashion — as has occurred with other kinds of computers.

As with classical, non-quantum computers, some industries will benefit disproportionately from the use of, and leverage provided by, more advanced computational power. The industries most likely to benefit are those that have massive amounts of data that need to be analyzed.

Because of its data-heavy nature, finance will likely be one of those fields that benefits early and asymmetrically from the ability to use quantum computers to analyze financial markets. This will impact agricultural commodity markets, trading, and price dynamics. Plus, quantum computing could be critical for weather forecasting and other agricultural activities I discuss later in this chapter.

The biggest hurdle to quantum computing is that it requires both a software and a hardware change. In fact, right now, one of the biggest physical challenges to true quantum computing is temperature, because a true quantum computer needs to operate at near absolute zero temperatures.

That means big physical and material science challenges lie ahead for quantum.

Also, it doesn't mean that you will someday have a giant quantum computer in your house that is being cooled to near absolute zero temperatures. It seems more likely that quantum computing processors would likely be accessed remotely, on a quantum as a service (QaaS) platform framework like many other cloud-based technologies and applications.

Of course, there is also the potential that you could have on-site room temperature coprocessors that utilize emulated or photonic quantum computer technologies. But those would still be different from true quantum computing.

In the highest-value use cases, it's critical to remember that quantum is likely to be not just useful — but also necessary. After all, data is being created at a breakneck pace — especially following COVID-19 digital transformations *en masse*. As data collection increases, more analytics and a better way to break down the meaning of the numbers will be critical.

An exponential pace of data creation, especially with the expansion of IoT connectivity, will necessitate a computational step change to derive meaningful implications from ever-larger data sets. In other words, quantum computing may become required for advanced predictive analytics and artificial intelligence. In the very near future, quantum computing is likely to be more than just nice to have, it could become a conditio sine qua non for data analysis.

Quantum computing also could help us push toward scientific and medical insights that are just too difficult to see with panel data that stretches the limits of current computational and statistical processing powers. There could also be some critical scientific insights that support agricultural production and supply chain activities.

Unfortunately, with a quantum computer, you need to ask the right questions. Quantum computing does not operate in a deterministic non-probabilistic fashion. Quantum computing is a non-deterministic, probabilistic kind of process. This means that if you feed it the data, it will spit out the least wrong answer. But if we can do that right, quantum could be a gold mine for data research, business activities, agricultural and supply chain optimization, scientific research, and society at large.

But quantum is not without its risks for the world. Quantum could present catastrophic risks for blockchain technology, cryptocurrencies, and all kinds of encryption technology. Given its non-deterministic probabilistic analysis, the highest-potential use case for quantum computing is widely expected to be in breaking encryptions and rendering all current forms of cybersecurity useless and presenting myriad threats.

Aside from the cybersecurity risks that may impact all businesses, governments, and individuals, most people may not directly notice the changes that quantum computing brings. After all, most retail computer users do not understand semiconductors. For that vast swath of people, quantum computing developments may go unnoticed.

Industries Likely to Benefit

As the Chairman of The Futurist Institute, I directed a 2018 analysis and study of corporate and commercial opportunities in quantum computing across industries. The overview is in Figure 12-1. We identified a number of industries likely to benefit from the computational power promised by quantum computing. Most of these industries deal with large amounts of data.

The industries we have identified include the following: finance, government, transport and logistics (e.g., e-commerce), energy, healthcare, and agriculture. Although agriculture is not likely to be the most impacted field by quantum computing, finance and supply aspects of food and ag could be greatly impacted.

Figure 12-1: Quantum Computing Assessment by Sector

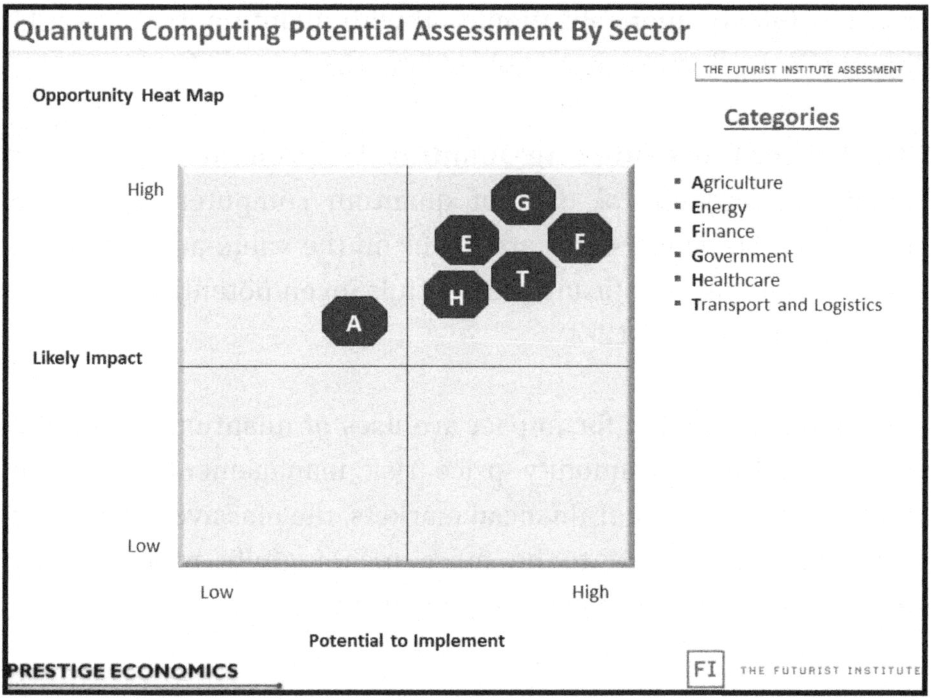

A number of the use cases are purely theoretical as a universal quantum computer with general-purpose applicability and integration with classical computing at a commercial scale does not yet exist. Also, for the purposes of our analysis, we excluded hard science, engineering, and mathematic use cases even though these could benefit from quantum computing, with massive research and scientific implications. We excluded these from our analysis because they may take longer to impact professionals, businesses, and the economy at large.

For the main industries we considered, we paid close attention to the already-existing use cases for predictive data analytics, machine learning, and artificial intelligence solutions. We concluded that it seemed reasonable that quantum computing could take these efforts to the next level if and when quantum reaches full commercialization. And with quantum, real AI could become a much more attainable technology.

The Futurist Institute's Ag Quantum Assessment

Food and agricultural uses of quantum computers are quite disbursed. Some use cases are lower on the value-add list, while others could be significantly impactful. Seven potential use cases are shown in Figure 12-2.

At the top of the list for impact are uses of quantum computing for agricultural commodity price risk management and price forecasting. As in other financial markets, the massive amounts of price data are likely to be more meaningfully processed by quantum computers than by current systems.

But the quantum computing value-add for food and agriculture doesn't stop with financial market risk management and prediction value.

There are also operational use cases for quantum computing in agricultural industries. Quantum could add value for planting and harvesting optimization. This could be integrated with market forecasts, anticipatory global demand, and financial rewards. There may even be the potential to use quantum computing to optimize food distribution to maximize quality and limit waste.

Plus, there are several other areas where quantum computing could be implemented, although these could see lower probabilities of implementation with more modest impacts.

Figure 12-2: Food and Ag Quantum Computing Potential

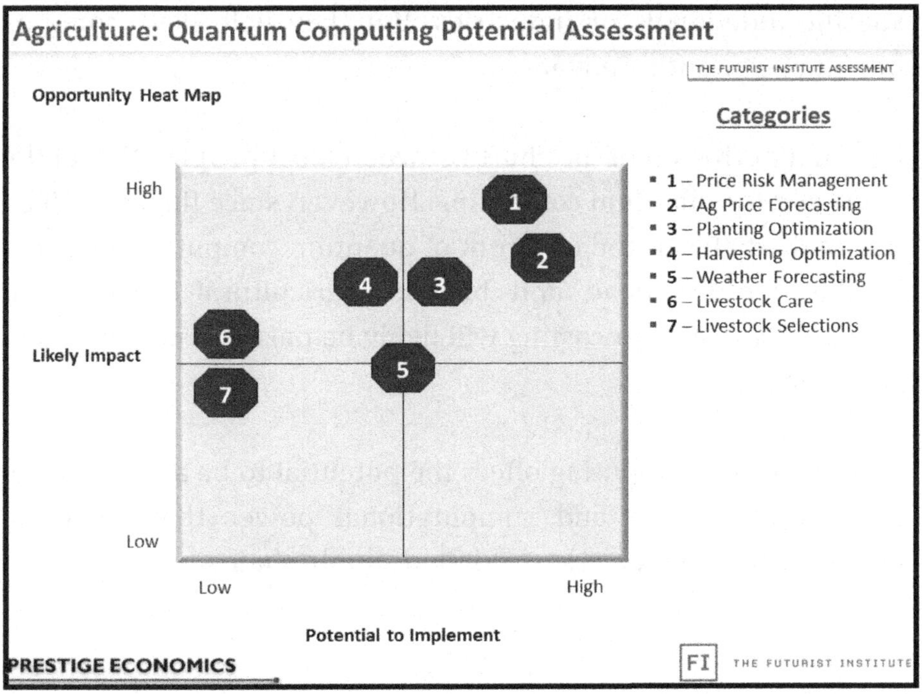

Weather forecasting, livestock care, and livestock selections round out that list. These uses may have lower value propositions, but quantum computing could theoretically be impactful at some level.

Threshold of Need

For most companies, near-term investing in quantum computing or installing quantum computers will not be a necessity — even if it were an option. The juice might just not be worth the squeeze, and it might be complete overkill. There is just not a high enough threshold of need.

Plus, for now, there is neither hardware nor software at a commercial level, let alone a very easy-to-use corporate interface or any quantum-reliant apps, that caters to the needs of any specific individuals or industries. But they will likely come as quantum computing advances.

Ag is not at the top of the short list when most people think of the value add for quantum computing. However, since finance will be one of the fields at the forefront of quantum computing adoption in the decade ahead and beyond, agricultural market and commodity price forecasting will likely be part and parcel of that adoption.

Plus, quantum computing offers the potential to be a step change in data processing and computational power that could be impactful for ag operations and food supply chains.

As I noted in my book *Quantum: Computing Nouveau*, quantum computing will be a new way of computing that will be critical for fields with massive amounts of data. Yet for many people the developments may largely go unnoticed. Nevertheless, it is a technology that is likely to impact the future of food and agriculture in the decade ahead and beyond.

PHYSICAL TECHNOLOGIES

CHAPTER 13

AUTOMATION

In January 2020, I led a project looking at the future of supply chain and e-commerce. An automation solutions client had come to me because their customers were skeptical about the future development and advancement of e-commerce. I was shocked. Honestly, I couldn't believe that people had thought that the prospects for e-commerce and automation had already been fully played out.

It didn't make any sense to me that this was a concern out there.

So, I pulled together research and crafted forecasts. Since I have worked in the e-commerce, material handling, and supply chain world for over five years, I've gotten pretty deep into some of the dynamics and technologies that will shape the future of our economy. In fact, automation, robots, robotics, and AI technologies are always the stars of The Futurist Institute's annual edition of *The Robot and Automation Almanac*. The result of the project was a very aggressive set of forecasts for e-commerce. At least that's how they looked in January 2020.

The most recent e-commerce data following the COVID-19 shutdown indicate that we are now well ahead of our heretofore aggressive forecast toward automated e-commerce.

But there is still more to come. And with increased levels of e-commerce, the need for automation will be greater than ever.

There is a significant and rising need for automated solutions to take goods out of factories, warehouses, and distribution centers and to bring them to consumers. This also includes last mile issues, which will become increasingly important in the decade ahead.

Automated tech has been deployed across the supply chain inside warehouses and distribution centers. This has been done, it will be done further, and it will continue to be a highly effective means of fulfilling the promise of e-commerce. But it is only one piece of the automation puzzle.

On the delivery side of the supply chain, we are likely to see drones, self-driving vehicles, and other automated means of transport deliver industrial, commercial, or consumer goods. This could include flocks of large industrial drones working together to deliver heavy industrial equipment to remote sites as well as performing agricultural tasks in robotic flocks of drones.

All of this is to say that the agricultural and food supply chains will require more automation — especially for repetitive, predictive, dangerous, and physically taxing jobs.

We are likely to see physical robotic proliferation across the supply chain — especially as software and hardware improve and costs fall. But we are also likely to see companies more aggressively deploy digital automated solutions as part of a cohort of technologies lumped together under the hypernym robotic process automation.

Disruptive Impacts on Supply Chains

As with the other areas of development, automation technology presents mostly upside potential — but it still comes with some downside risks. On the upside, automation increases efficiencies and the speed of goods movement.

It also increases the ability to manufacture or deliver goods that were previously complicated or expensive to deliver. And automation can also ensure more stable and predictive delivery of industrial, commercial, and retail goods.

In the wake of COVID-19, people received food deliveries from people. But many people would likely have rather had robots bringing them food, to reduce the risk of COVID-19 exposure.

Looking out a decade, it is likely that people could be significantly removed from the agricultural and food supply chains. Between the mix of increased speed, convenience, and predictability, it should be clear that the increased deployment of automation is likely to reduce delivery costs and increase profitability while also increasing food safety — something that really was not at all on the radar before COVID.

In addition to the use of robots across the entire agricultural and food supply chain, there is also likely to be an increase in the sales of robot installations in the food and beverage industry. This would be another way that COVID-19 and any future pandemic contagion risks would fall.

Furthermore, using robots for these jobs could also increase operational stability. After all, robots working in restaurants do not get sick, need time off, have family emergencies, or take vacations. These dynamics have been fueling the increasing recent trend in robot sales shown in Figure 13-1. Furthermore, I expect these sales will continue in the decade ahead at a more rapid pace.

Figure 13-1: Food and Beverage Robots[1]

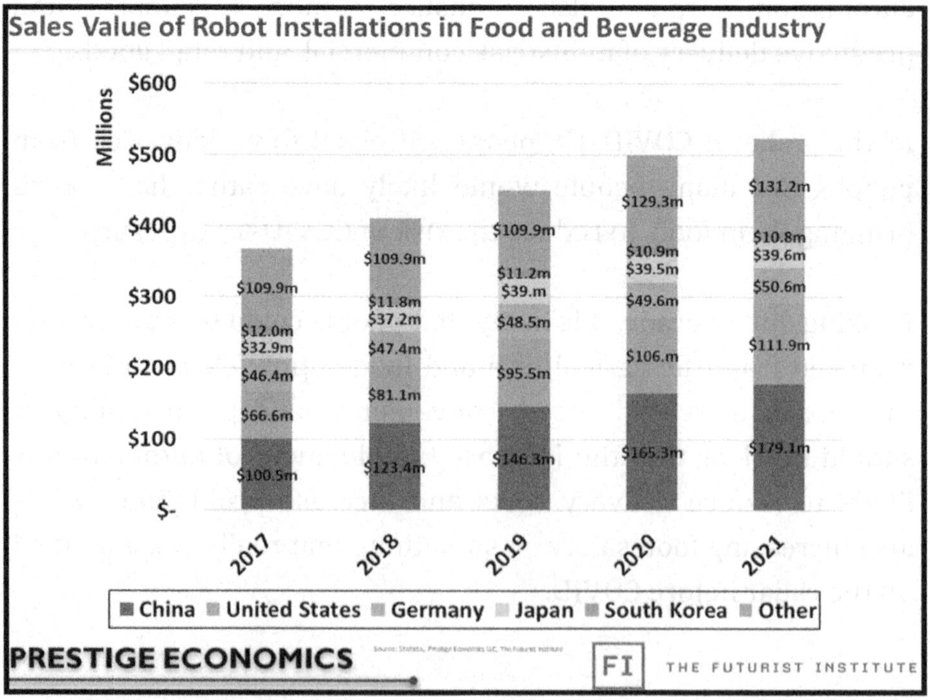

Despite all of the upside potential for the use of robots and automation in its many forms in agribusinesses and across food and agricultural supply chains, there are a couple of downside risks posed by automation as well.

Most importantly, not all companies will have the same capital to invest to make the transition from the information age to the automation age. And preparing for a transition to the automation age requires significant capital investment on the part of companies. This means that there will be winners and losers. Companies that make the right investments will be the ones that capture a greater potential of upside in the future. And the adoption of more rapid automation solutions is likely to have a compounding impact on profits.

We saw this dynamic during the COVID-19 pandemic, as companies that were able to provide goods through e-commerce channels were able to grow and profit. One other risk to consider is that between 2.2 and 3.1 million transportation jobs in the United States are likely to be lost due to automation. Other jobs lie ahead, but these losses could have a substantially negative impact on many individuals and society — even if public health outcomes might improve by removing people from the last mile delivery of food and agricultural products in a post-COVID world.

CHAPTER 14

OPERATIONAL DRONES

One of the most important physical technologies that will be broadly deployed in the next decade are likely to be industrial drones. These are likely to become parts of last mile solutions to meet rising e-commerce demands in the United States (and globally) — especially after the COVID-19 pandemic.

These autonomous self-driving and self-flying vehicles are likely to use a mix of radar and lidar to get goods to people. Industrial drones will also be used for remote operations, to repair infrastructure, support business operations generally, and engage in physical activities in agriculture and mining.

Agricultural work is hard work, which is why the use of drones and automated physical machinery is likely to be increasingly deployed in the decade ahead and beyond, whether we are talking about planting, harvesting, culling, or transporting agricultural goods, products, and food. Across much of the mid-stream of the U.S. supply chain, automated vehicles have already been broadly deployed. But that's only part of the chain.

It still leaves the upstream and downstream parts of the economy to be automated. Granted, the world outside warehousing and distribution centers (and other structures of operationally two-dimensional space) is much less predictable and unstable.

But the technology for automated vehicles and drones is improving rapidly. The use of automated vehicles in the supply chain has been driven by the reality that there simply aren't enough people to move all the goods through the supply chain as well as the fact that people cannot work with the same speed and have the same strength as automated solutions like drones and automated guidance vehicles.

A similar threshold of need is waiting for food and agriculture. Thinking back to the figures in Chapter 5, we need to find a way to produce 44 percent more calories in just 30 years. Clearly, putting people in fields with rakes and hoes isn't going to get us there. But automating more activities could result in efficiency gains and yields in a way that can help resolve the global hunt for calories in a way that just using people likely cannot.

In short, agriculture is likely to be increasingly automated in order to ensure that those four meals MI5 is worried about don't become an issue. I mean this only somewhat in jest. In truth, the reality is that food could become a higher priority — and maybe even more expensive — good, attracting additional capital to the space in the decades ahead in order to meet the needs of a growing populace of much wealthier global citizens.

At the time this book went to print, most people thought of flying drones as toys with entertainment or kitsch value. You know, something you might have bought at Brookstone as a holiday gift before COVID-19 happened.

Of course, for supply chain purposes, the greatest future kind of drone isn't going to be 12 inches wide and carry a digital camera around as it follows you to record every second of your next European vacation for posterity on Instagram and Snapchat.

Those kinds of drones will grow in use as well. But those aren't the kinds of drones that will disrupt supply chains and work to solve last mile challenges. The better comparison for drones that will have big economic impact would be flying military drones that are adapted to have a dual use purpose for transporting industrial equipment to remote locations.

Retail consumer demand is often assumed to be the most critical use case for drones, but these kind of dual use industrial drones with 5 to 10 meter wingspans (or more) will likely add far more economic value. Remote operations that require large equipment deliveries are more likely to be serviced by drones than someone's online order for toothpaste.

This is just as true for mining as it is for logging, fishing, farming, or other agricultural activities. Industrial drones could allow businesses to access previously impractical and remote acreage for farming purposes. And those drones could help ensure security of food and agricultural supply chains in the way that a shepherd and his sheepdog cannot.

Of course, we are also likely to increasingly see flying drones used for e-commerce delivery. These are likely to be both in the air and on the road in the decade ahead and beyond. Plus, they may be used on their own or in conjunction with on-road distribution center autonomous trucks or airborne drone motherships.

On-road delivery bots may be deployed in densely populated and relatively flat urban environments first. But they may be quickly surpassed by full-size autonomous delivery vehicles. And their biggest challenge will be the notion that "if man can make it, man can break it." And these bots may be subject to high levels of theft and mischief as they struggle to achieve broad adoption — especially in sprawling suburbs and areas with less-than-flat topographical layouts.

Flying drones, however, could find increased use in the already highly-automated midstream of the supply chain where they could be used in warehouse and distribution facilities for picking and pulling.

Through mid-2020, most of the drones operating in warehouse and distribution centers were on wheels and operated in two-dimensional space. But drones could go airborne in these facilities as e-commerce demands rise. Similarly, flying drones could also be deployed in hydroponic facilities and other parts of the emerging agritech world if industrial designs of these facilities strive for a contactless food approach.

As with other disruptive technologies ahead, there will be challenges for drones. One of the biggest risks will be that drones may be hackable and regulations may prohibit industrial and retail last mile drone development. This is especially a risk if roads and skies become very crowded with drones that appear to be everywhere.

In the end, the biggest supply chain value add for drones may not be in last mile solutions in the next 10 years. But it does seem likely that flying drones could have a material impact in extending the reach of global supply chains by increasing access to locations and markets that were previously exceptionally isolated. This could be a very meaningful development for food and agricultural supply chains in the decade ahead and beyond.

SUSTAINABILITY AND SHORTAGES

CHAPTER 15

ESG AND SUSTAINABILITY

One of the big trends for the future of food and agriculture will be the recent and future push to incorporate environmental, social, and governance (ESG) goals along with sustainability. After all, the future of food and agriculture are likely to be replete with both environmental as well as social challenges.

The push for ESG goals is a trend that has been underway for some years, and in the wake of the COVID-19 pandemic and shutdown many companies have drastically reduced their carbon footprints. People are traveling less and commuting less, and there may be incentives for companies to show ESG progress that could keep many people working from home — and keep travel depressed for many years after the greatest COVID-19 fears have faded.

These developments are positive, but there are going to be big challenges in the decades ahead as the global population rises, global caloric needs rise, and global protein caloric needs rise. The ESG element is one of the critical reasons why lab-grown meat and hydroponics could be critical technologies.

As we think about environmental sustainability, the limits of global agriculture could be pushed to the brink in the decades ahead by global demand. At the same time, food deserts, where people cannot afford fresh food at reasonable prices are likely to become more ubiquitous, presenting social challenges.

In 2018, some of the top areas where there were activist investors pushing for goals and initiatives included climate change (19%), sustainability (13%), other environmental (7%), and political activity (19%).[1] You can see the breakdown in Figure 15-1. If we consider sustainability a type of environmental initiative along with climate change and other environmental resolutions, we see that 39% — a plurality — of all activist investor resolutions filed in 2018 were environment related.

Figure 15-1: Type of Activist Resolutions Filed in 2018[2]

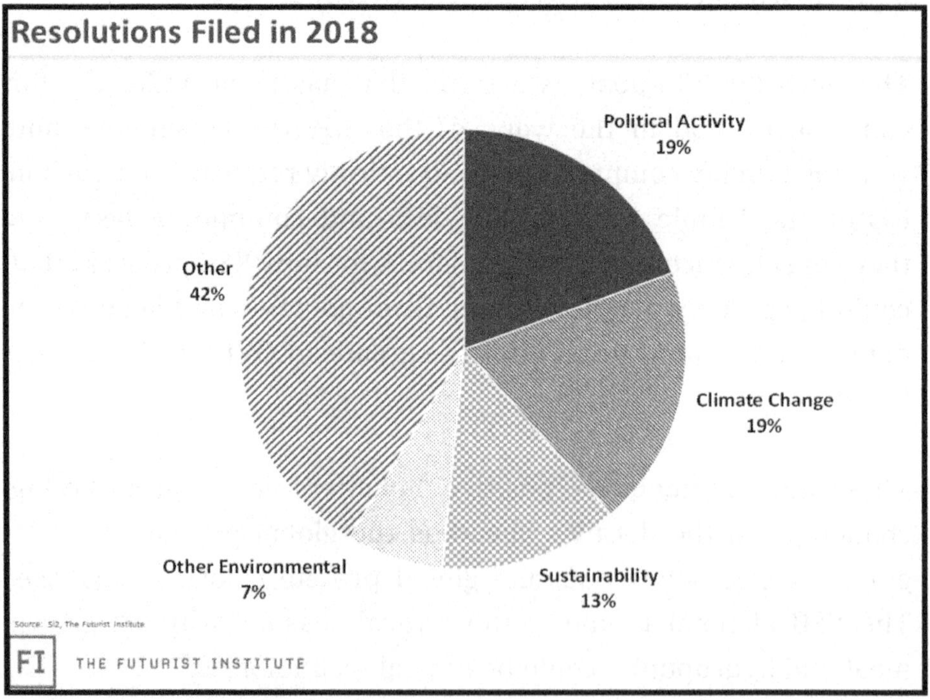

Figure 15-2: Global Companies Subject to Activist Demands[3]

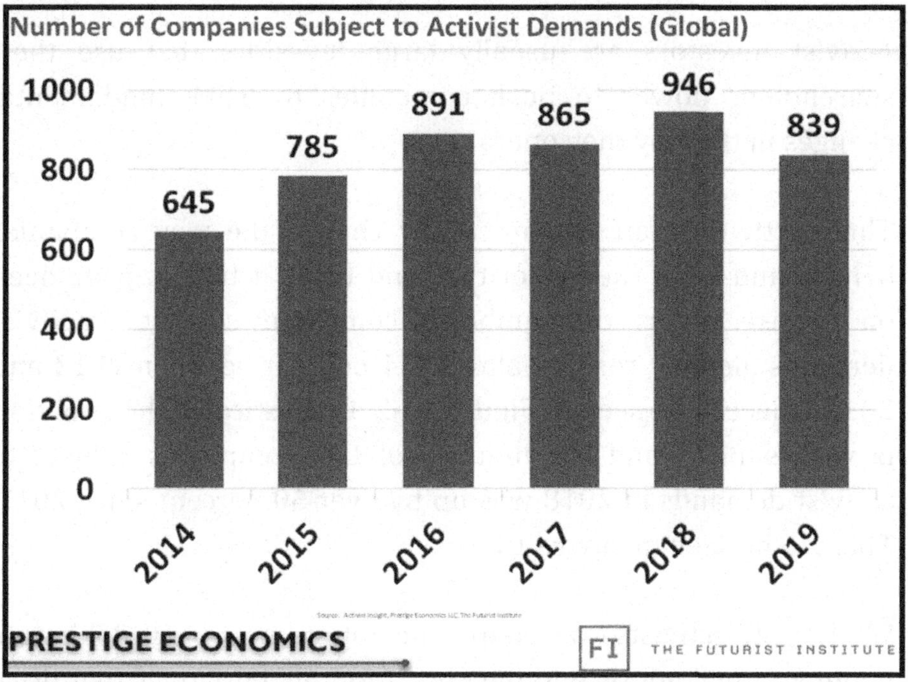

Figure 15-3: U.S. Companies Subject to Activist Demands[4]

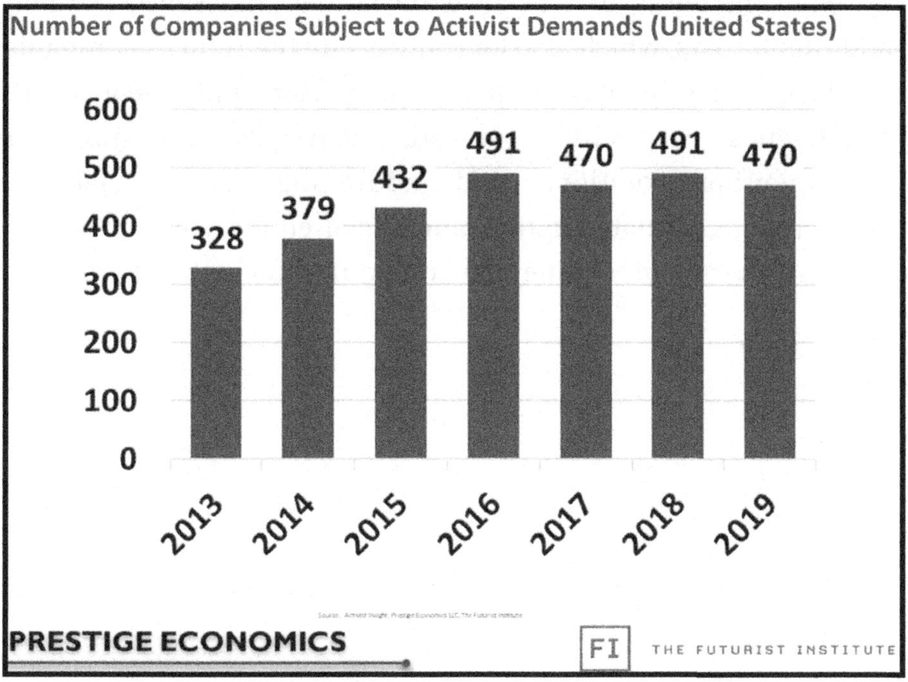

Activist investors are usually large investors that use their shareholder power to push companies to make fundamental changes in the way they operate.

Their activities can fundamentally change the way companies behave and what they prioritize, and those activities have been on the rise. In fact, the number of companies subject to activist demands globally rose by almost 54 percent between 2013 and 2018. This can be seen in Figure 15-2. Domestically, the situation is very similar, and the number of U.S. companies subject to activist demands in 2018 was up by over 50 percent since 2013. This can be seen in Figure 15-3.

While ESG activists can shape the future in a very deliberate manner, it will not be without cost. After all, initiatives that price in environmental or social externalities inherently embody additional costs, which also logically erode the returns companies produce. Therein lies a potential poison pill, because the effectiveness of activist investor demands will hinge on implementing solutions and addressing issues like the environment, climate change, and social equity without eroding so much shareholder value that companies fail in the process.

CHAPTER 16

MOVE TO "CLEAN FOOD"

In addition to the push for clean energy and sustainability, there has also been a rising trend toward food that meets some level of "clean" guidelines. This wide range of food and beverage grouping includes alcohol-free beer, wine, and spirits as well as non-genetically modified organisms (non-GMOs), organic, vegan, halal, and kosher foods. Furthermore, these trends are generally poised to continue and accelerate.

In Figure 16-1, you can see how organic food demand in the United States increased by around 250 percent in just a decade. Like other food categories listed here, organic food buyers are also willing to pay a premium for food that is clean.[1]

Naturally, this is very different than the notion of ritually clean foods, when it comes to kosher or halal foods, for example. But an increasing global Muslim population will likely support continued growth in the halal market. Interestingly, in some countries, like the United States, kosher food is seen as a viable substitute when halal foods cannot be procured.

For that reason, kosher food sales could also grow in the decade ahead and beyond in the United States and outside countries where Muslims are a majority of the population.

Another impact of the growth in the global Muslim population is the demand for non-alcoholic beverages, because alcohol is viewed as *haram* — or religiously prohibited in Islam.

In addition to people seeking a healthy diet, observant Muslims are likely to be a key target demographic for this product. In many ways that makes non-alcoholic beverages a market with a critical overlap, for both increasingly health conscious consumers as well as a vital growing demographic.

Figure 16-1: U.S. Organic Food Sales[2]

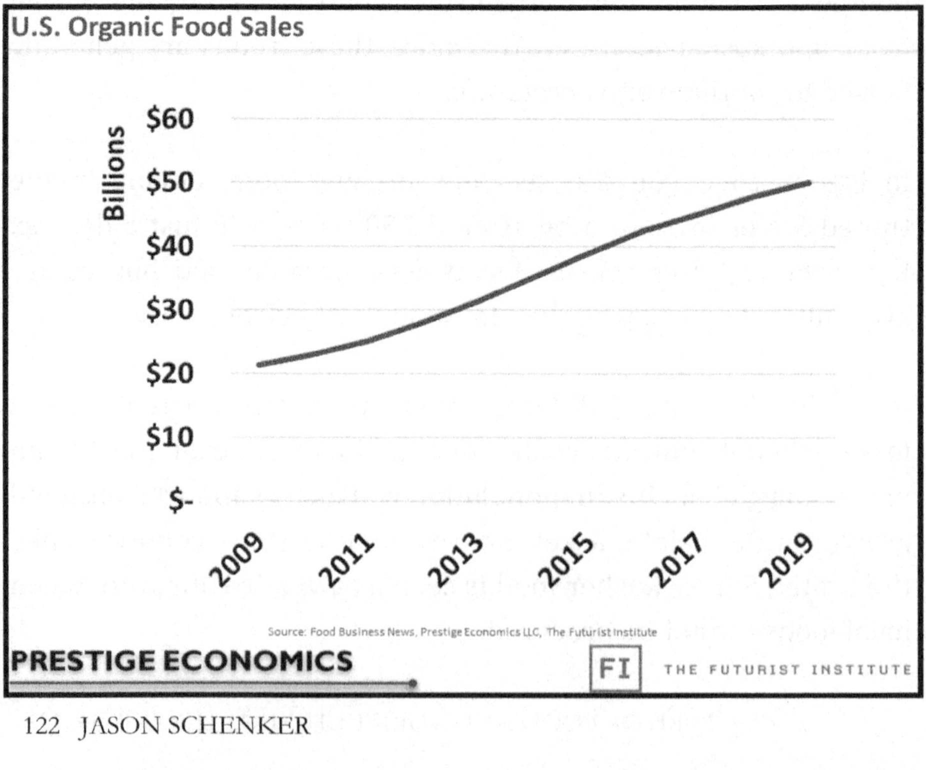

As an interesting twist, lab grown meat may be kosher — even if it is pork.[3] Whether lab grown pork actually gets categorized as *treyf* (non-kosher meat) — or haram — remains to be seen, but manufactured meat will present some challenges for religious authorities.

Another critical issue in the decade ahead that will pose issues for global authorities — albeit of a secular nature — will be genetically modified organisms (GMOs). The debate is likely to rage on with increasing ferocity as the need for more calories drives technological genetic modifications and innovations in the fields of food and agriculture that may involve genetic modifications.

Figure 16-2: Global Muslim Population[4]

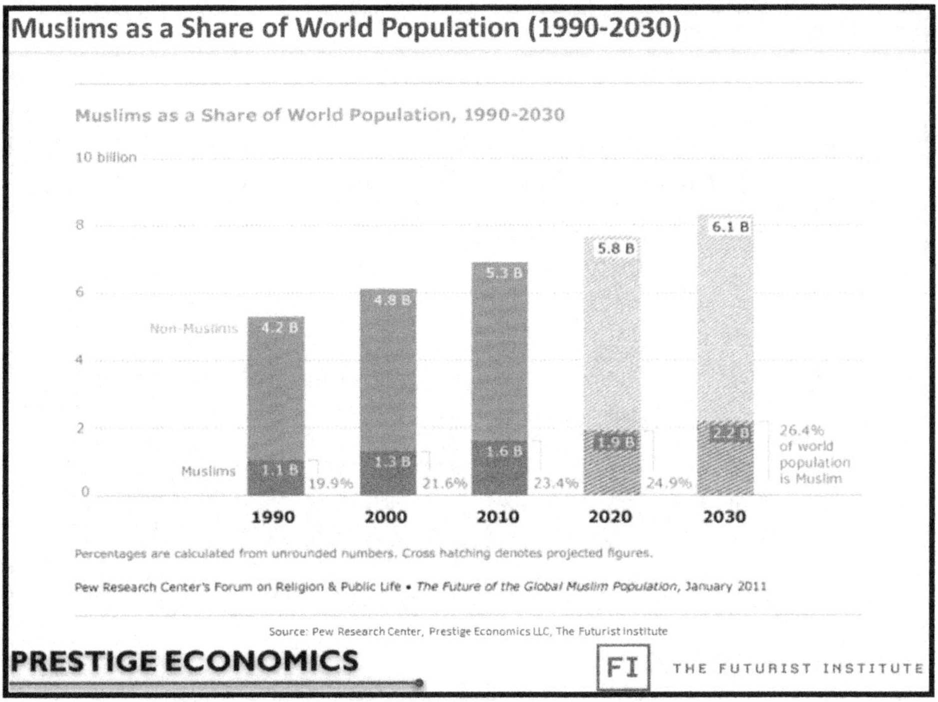

Even though GMOs will be enlisted to meet global needs for calories, a pricing premium is likely to be placed on non-GMO products. I will discuss the GMO dynamics further in Chapter 19.

The final specialty markets we should consider are milk and meat alternatives, which function as vegan food options. These products have gained significant market share, as you can see in the milk alternatives graph in Figure 16-3. These two vegan products have seen tremendous sales growth in recent years. Moreover, product demand for these milk and meat alternatives — as well as other vegan products — is likely to continue growing in the future as well.

Figure 16-3: Milk Alternatives[5]

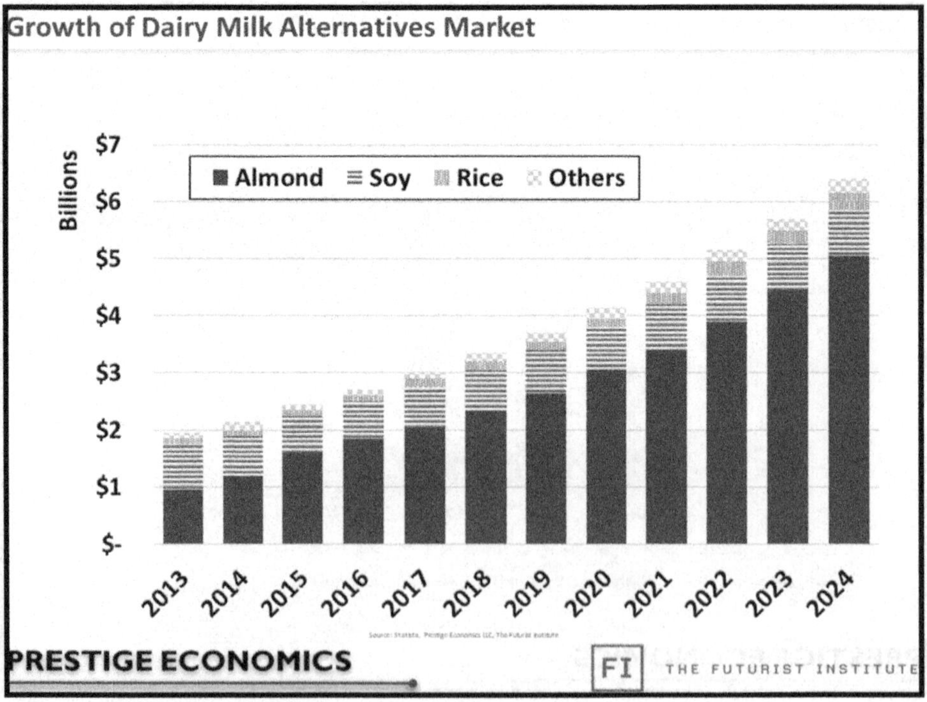

Outlook for Specialty Markets

Specialty markets of all kinds are poised to grow overall — especially as global population and global wealth increase. Furthermore, it seems likely that the profitability of specialty products could continue to exceed non-specialty foods, especially as the preference for foods that meet certain dietary restrictions continues to rise.

Figure 16-4: Meat Alternatives[6]

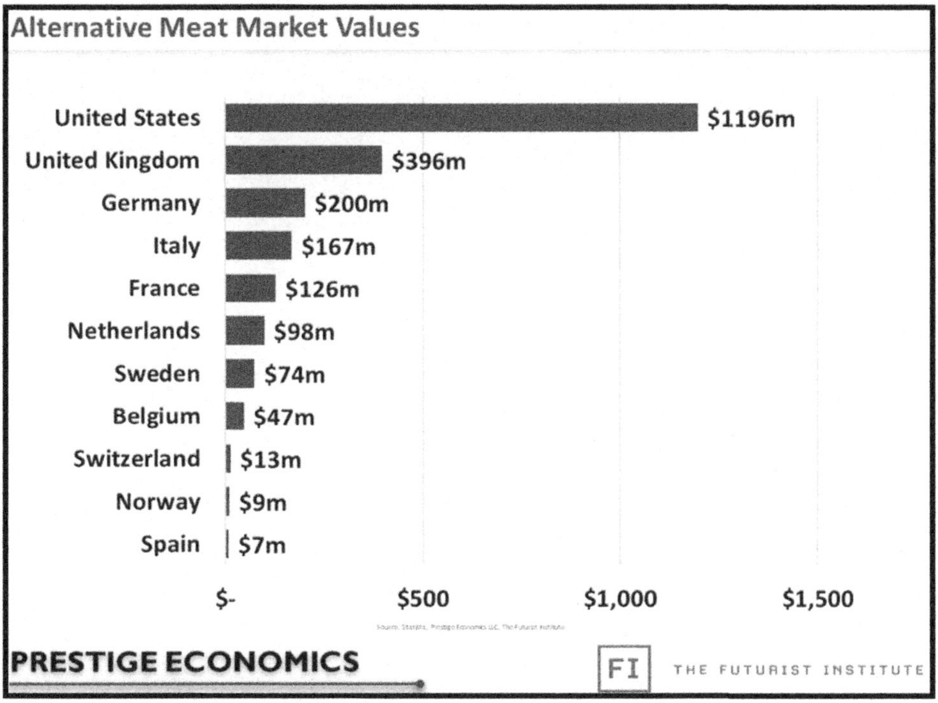

CHAPTER 17

WASTE-TO-ENERGY

One area where the future of agriculture and the future of energy overlap is waste-to-energy, which is the creation of power by burning solid municipal waste. While there is a big focus in agricultural research on CO_2 output and water use, there is another area where a rising global population can have a significant environmental impact — and that's in landfills.

Waste-to-energy is not quite a renewable energy and it isn't exactly a clean form of energy, but burning solid municipal waste to generate energy can reduce the total overall environmental impact of solid waste that would otherwise sit in landfills or might be dumped into the sea.

In 2015, 262 million tons of solid waste were burned in the United States. The type of waste was distributed among a number of different waste categories, but paper was a plurality of the waste material, at almost 26 percent of U.S. waste.[1] The next most important solid waste categories were food waste, at over 15 percent, and yard trimmings and plastics, at just over 13 percent of total solid waste each.

This means that rather than throwing away paper, food waste, and other refuse into landfills, it could be used to generate energy and offset fossil fuel power generation.

In the United States, waste-to-energy power generation is at a relatively low level, and it has remained relatively unchanged since the mid-1990s, as you can see in Figure 17-1.

Compared to a number of other countries, the United States burns a relatively low percentage of its total municipal solid waste. This is due, in part, to a few relatively regional concentrations of waste-to-energy facilities in the United States.

Figure 17-1: Solid Waste Power Generation Capacity[2]

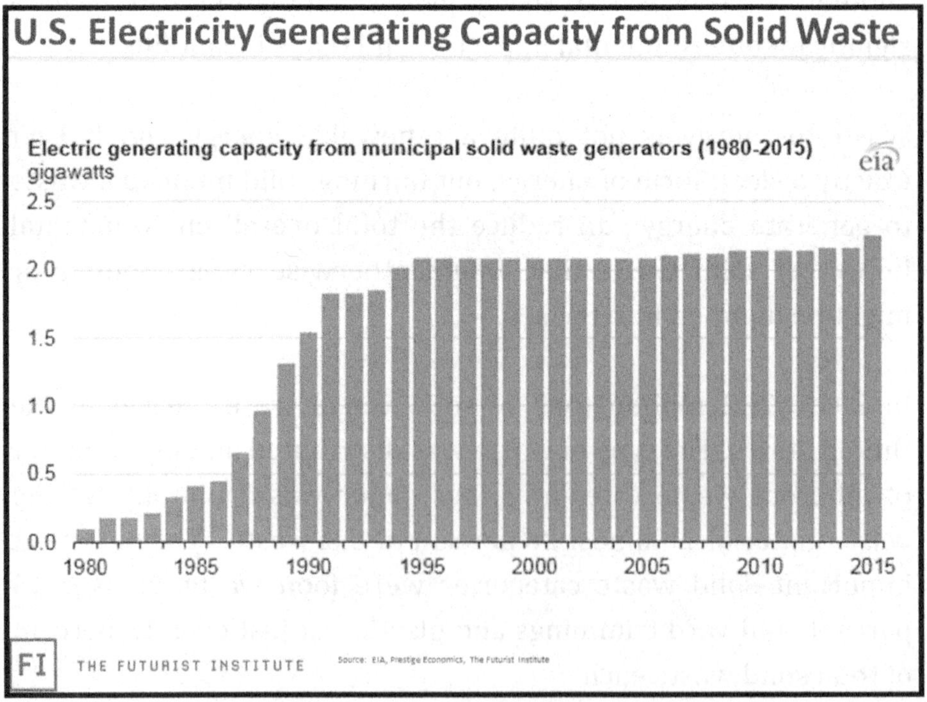

In some countries, however, waste-to-energy power generation is significant. In Figure 17-2, you can see comparative rates of burning total municipal solid waste for energy across countries. In Japan, for example, 68 percent of total municipal solid waste is burned with energy recovery. Germany burns 25 percent of its total municipal solid waste. But in the United States, only about 13 percent of total municipal solid waste is burned for energy.[3]

On a global basis, power generation from solid waste is likely to increase. This is especially true as emerging economies grow and generate more solid waste. As the population of Planet Earth rises by over 2 billion in the next 30 years, land will become a more valuable asset — and it will be priced at a premium.

Figure 17-2: Solid Waste Burned by Country[4]

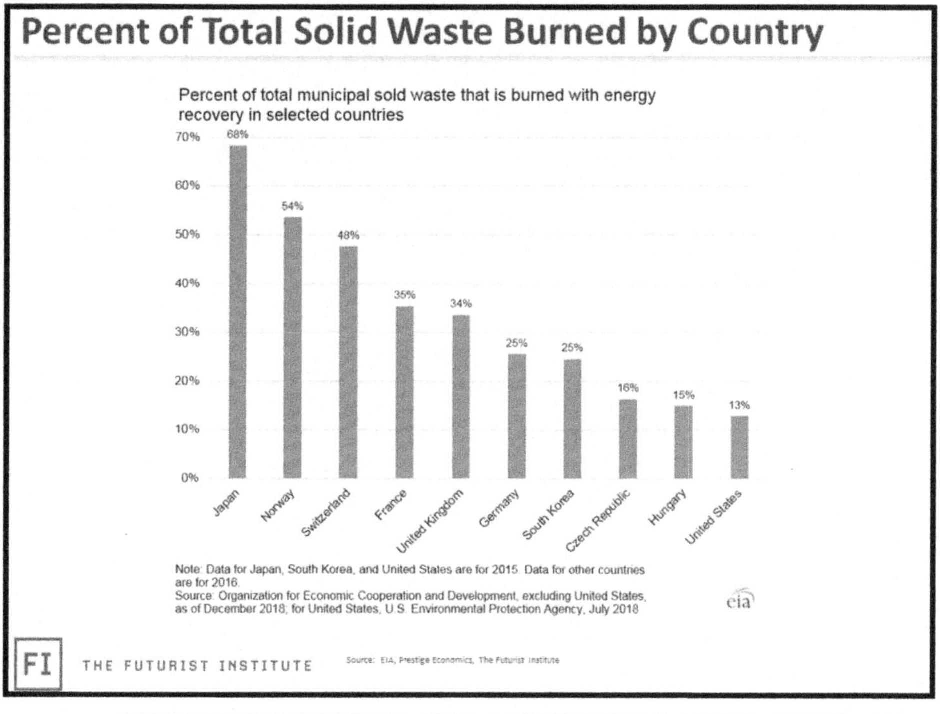

CHAPTER 18

ETHANOL

Looking out into a future where global calories will be needed in significantly greater numbers, food that becomes fuel is likely to find a better and higher use as food rather than being converted to fuel. While biodiesel is often made from used vegetable oil, ethanol is usually produced from virgin corn or sugar, quite literally turning food into a kind of alcohol that your car runs on.

Ethanol subsidies were created to support corn farmers in the United States, and they have made ethanol an attractive fuel to create. But if the global population increases as expected, there will be less need for corn subsidies in a future where food experiences greater demand. Furthermore, the use of sugar as a fuel — although common in Brazil — is also likely to fall as global calorie demands rise.

But for now, ethanol vehicles are a majority of the U.S. alternative fuel vehicles. In fact, 79 percent of all U.S. alternative fuel vehicles were ethanol vehicles in 2017.[1] But the outlook for food demand — as well as the rise of electric vehicles — does not favor ethanol vehicles.

Ethanol vehicles are set to lose alternative vehicle market share to electric vehicles through 2050. But electric vehicles will only equal ethanol vehicles as a percent of alternative vehicles by 2037, according to EIA forecasts, which can be seen in Figure 18-1.[2]

Even as late as 2050, 16 million ethanol vehicles are expected be 23 percent of 73 million U.S. alternative vehicles, with an overall total fleet size of 295 million vehicles. Ethanol is poised to remain important as an additive in the future of energy to stretch oil molecules,[3] but global food demand is likely to put the kybosh on any chance at broad-based ethanol use or fleet expansion. Simply put, the ingredients will likely be more valuable as food than fuel.

Figure 18-1: Ethanol and EV Proportion of U.S. Alternatives[4]

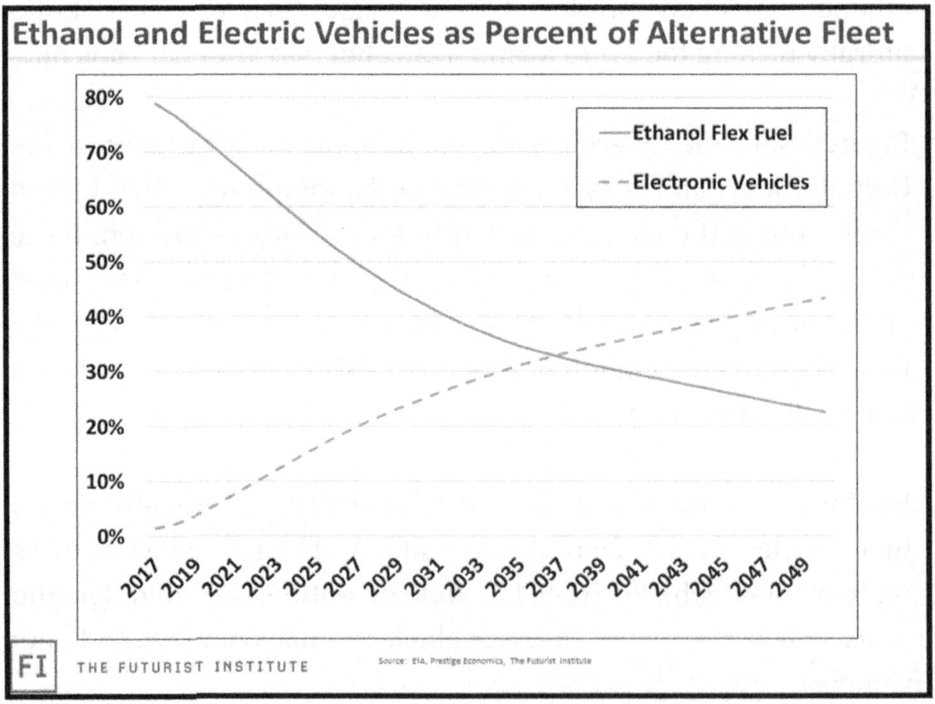

DISPUTED TRENDS

CHAPTER 19

FOOD EXTINCTION RISKS AND GMOS

One of the most hotly disputed trends in the world of food and agriculture are genetically modified organisms (GMOs). Against a backdrop of rising global temperatures and climate change, some foodstuffs could see drastic changes in their availability as they are subjected to overharvesting and reduced production due to sensitive climate requirements. GMOs could be a way to save them from extinction. But even making that assertion would be read as quite scandalous to some.

Let's look at two of the most important agricultural commodities that are at risk of extinction in the decades ahead: chocolate and coffee.

Both of these crops require very specific conditions to flourish, and if global temperatures keep rising, chocolate (i.e., cacao) could go extinct in 40 years.[1] For coffee, the prospects are similarly dire and 60 percent of wild coffee could go extinct in the decades ahead.[2]

In the same way that fish are likely to be increasingly raised at fisheries or in hydroponic closed cycle facilities to offset potential overfishing extinctions in the wild, averting wild coffee and chocolate shortages, as well as outright extinction risks, will require significant efforts.

In this case, one potential solution could involve genetically modifying chocolate and coffee plants. Despite being often maligned for its risks, GMO solutions could prove critical for preserving and sustaining these very temperature–sensitive and climate-sensitive foodstuffs, if they are to survive in conditions that could be less suitable to supporting their traditional growth and harvesting.

According to WebMD, back in 2016 GMOs were already found in 80 percent of packaged foods in the United States, and they are likely to proliferate with greater frequency in the decade ahead. And the science appears to safe, but a debate rages on.[3]

For a full discussion on GMOs, I recommend consulting the U.S. FDA website here — https://www.fda.gov/food/agricultural-biotechnology/science-and-history-gmos-and-other-food-modification-processes.[4]

Some countries are likely to continue to push back against GMOs and some consumers will eschew their consumption and happily pay a premium for non-GMO foodstuffs, which is something I mentioned back in Chapter 16. But as the world pushes to meet rising global calorie demands, GMOs are likely to be a critical part of the solution to improve yields and potentially address other issues like removing CO_2 from the atmosphere.

CHAPTER 20

THE FUTURE OF WATER

Water access is one of the biggest problems the global economy, society, politicians, and agricultural businesses will face in the decade ahead and beyond. The challenges associated with water risks were in place long before the COVID-19 pandemic, shutdown, and recession. But it should now be even clearer following the social unrest associated with COVID-19 that if the supply chain of basic necessities breaks down, even the most advanced OECD economies could face structural systemic risks.

In other words, ensuring access to basic necessities like water is critical — and will become even more critical in the years ahead, especially if climate change results in increased droughts and shortages. Plus, the current state of play is not great.

According to a 2019 report from the World Health Organization, there are currently 785 million people who lack "a basic drinking water service." Moreover, at least 2 billion people currently lack access to safe drinking water at home and by 2025 "over half of the world's population will reside in water-stressed areas."[1]

That's just five years from now — and the risks through 2050 are much greater. In some regions where water shortages are exacerbated by climate and population dynamics, resource wars over water are not unforeseeable risks. This is likely to be a particularly acute risk in places like the Middle East that are already deeply entrenched in multi-lateral proxy wars.

Agriculture is a critical part of the future of water, because global water demand has been driven predominantly since 1900 by agriculture. In Figure 20-1, you can see a United Nations graph showing historical global water withdrawals. Although all categories have increased in the past 40 years, agriculture has been and remains the lion's share of water use globally.

Figure 20-1: Global Water Withdrawals Since 1900[2]

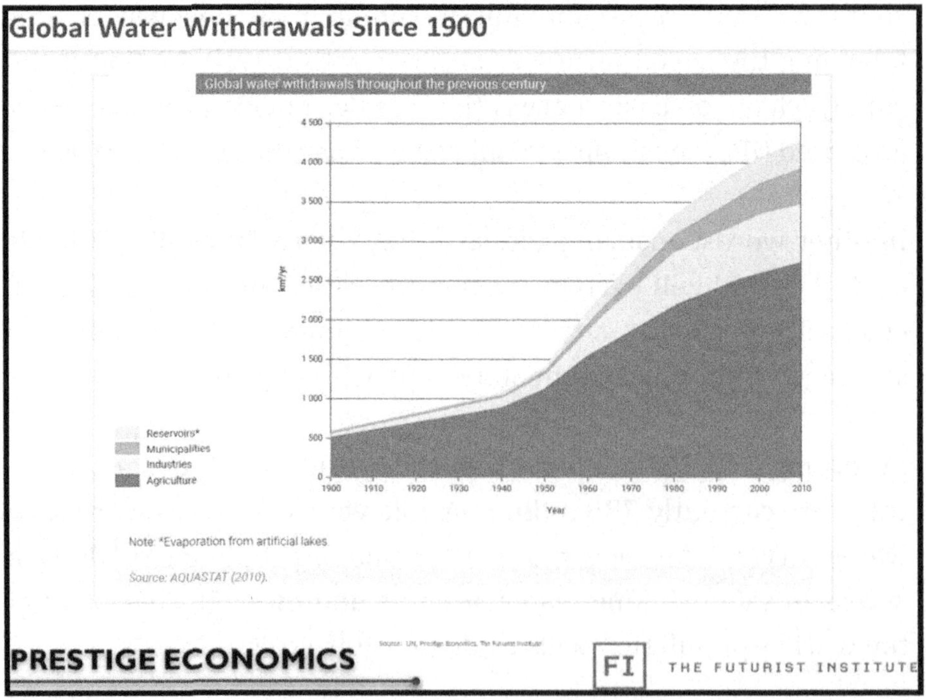

Agricultural uses of water have dominated the past and the present — and they will likely dominate the future. But one of the biggest areas for water demand growth through 2050 is likely to be industrial demand. Globally, industrial water demand is expected to increase by 165 percent.[3]

This dynamic is shown in Figure 20-2, where you can also clearly see that the biggest growth rates will mirror where population growth and emerging market middle class wealth will dominate the decades through 2050 — in Africa and Asia. In sum, the global share of industrial water use by 2050 is also expected to increase to 24 percent from 18 percent in 2010. This means that even though calories will be critical, industry needs water too.

Figure 20-2: Industrial Water Demand in 2010 and 2050[4]

Industrial Water Demand, 2010 and 2050

Industrial water demand, by continent, 2010 and 2050 (*middle of the road* scenario)

	2010 (km^3/yr)	Share of continent's total water use	2050 (km^3/yr)	Share of continent's total water use	Change rate (2050 as % of 2010)
Africa	18	8%	64	18%	353%
Asia	316	10%	760	19%	240%
North and Central America	229	35%	182	27%	80%
South America	31	19%	47	21%	153%
Europe	241	54%	325	58%	135%
Oceania	2	5%	3	7%	144%
World	**838**	**18%**	**1381**	**24%**	**165%**

Source: Adapted from Burek et al. (2016, table 4–10, p. 62).

Source: UN, Prestige Economics, The Futurist Institute

PRESTIGE ECONOMICS

FI THE FUTURIST INSTITUTE

Of course, there's only so much fresh potable water on the planet, which is why IEA forecasts shared in a 2020 UN report also reflect expectations of a significant increase in water transfer and desalinization.[5] This underscores the risk that water could become a traded and transported commodity that requires significant energy to desalinate and transfer between locations.

The commerce and potential conflicts around water have long been underway, with many countries seeking to shore up access to fresh water in regions destined for conflict. Looking to the decade ahead, there is great potential for global conflict as the United States and China jockey for the pole position as part of a great power competition. As the risk of associated proxy wars increases, water access could be a key asset that brings countries into alignment on one side or the other in the event of a Second Cold War.

CHAPTER 21

TRADE WAR DYNAMICS

The United States delivers 44.4 percent of its agricultural exports to Asia, including 12.1 percent to China.[1] But the escalating great power competition between the United States and China has disrupted their symbiotic trade relationship. Elevated tensions as part of a trade war that started in 2018 have become more acute in the wake of the COVID-19 pandemic, shutdown, and recession, and the tensions between the U.S. and China could present much greater risks to finance, the global economy, and trade.

Most germane to our evaluation in this book, the trade war between China and the United States has already been significantly impacting some agricultural commodity flows, markets, and prices. Furthermore, if the tensions between these two "frenemy" nations become more severe in the future, it would present even greater risks to the free flow of goods and services between their economies, throughout their supply chains, and across the globe.

Even without a trade war escalation, a failed attempt to de-escalate the trade war could further significantly impact food and agricultural product flows, markets, and prices as well as global supply chains long after the COVID-19 recession has ended.

The COVID-19 pandemic has opened the aperture in many ways, revealing open secrets and overlooked risks — for society, the global economy, and national security. COVID-19 has also revealed the risks of overly lean supply chains as well as the downside of being dependent on a global supply chain for critical goods like medical supplies, medical devices, basic necessities, and personal protective equipment (PPE), like gloves and masks.

Figure 21-1: U.S. Ag Exports by Global Region[2]

U.S. Agricultural Exports by Region

	Share of World		Share of U.S.
Region	GDP	Population	Ag. Exports
North America	27.7	6.5	28.4
Emerging Markets	24.1	45	16
Europe & Central Asia	24.9	9.3	10.4
Asia & Oceania	34.0	55.2	44.4
Latin America	4.6	5.8	23.9
Middle East & N. Aftica	4.2	6.5	5.9
Sub-Saharan Africa	2.0	13.3	1.2

Source: USDA, Prestige Economics LLC, The Futurist Institute

PRESTIGE ECONOMICS

FI THE FUTURIST INSTITUTE

COVID-19 also exposed the fragile nature of social stability and national security when food and agricultural products became tougher to come by.

Looking Back to Look Ahead

The global tensions in the trade war between the United States and China that started in 2018 have only worsened. In fact, in the wake of COVID-19, there are even risks that we could be on our way to embarking on the Second Cold War.

Key trade risks between the U.S. and China contributed to a slowing of global growth in 2018 and 2019. This followed strong global growth in 2017, which was supported by strong global trade. But there is more at play between the United States and China than just tariffs. There is a struggle for global economic, political, and military hegemony.

This is why, in part, the U.S.-China trade conflict remains largely unresolved as of August 2020. Furthermore, there are rising risks that the conflict could continue and proliferate further. As that occurs, the downside risks to overall global economic growth will increase. I have previously noted that this may require bifurcated supply chains to serve U.S. and China end markets in the future — especially for high-tech goods.

One of the most critical questions now is, how will international relations between China and the United States change — and how could it create ripple effects in global supply chains, the global economy, and great power competition?

I recently tried to answer this question by presenting scenarios for the future of supply chain after COVID-19 in an unclassified Pentagon report.

You can download the full unclassified Pentagon report here:
http://www.futuristinstitute.org/pentagon-report/

As we consider the importance of trade for the future of food and agriculture, we need to recognize that global economic growth is highly sensitive to global trade. And there are broader global macroeconomic risks in a high-risk trade environment. If the conflict between these two great powers becomes tenser, as seems likely, growth and financial markets could come under pressure in the future.

Fortunately, however, for the food and ag outlook, even during the height of the Cold War, grains and other agricultural products flowed between the U.S.S.R. and the United States. This means that even in an escalated trade war environment, ag goods could still flow. The question is will they flow — or will those flows be more protractedly choked off in the conflict?

CHAPTER 22

THE RISE OF WEED

Although more frivolous than the global debates and conflicts over GMO, water, and trade, marijuana is a hot button issue that is very much a source of conflict in the United States. This may change, but it could still take time. Meanwhile, companies are moving forward building their businesses as best they can by focusing on the states where marijuana has been legalized.

The reason I have included this topic in this book is because the outlook for marijuana legalization has never been more likely in the United States. Furthermore, while a number of cities and states have decriminalized or legalized marijuana use in one form or another (i.e., as medical or recreational), the chances of seeing more legalizations are much greater after COVID. And it all comes down to the cash.

Following the COVID-19 pandemic, shutdown, and recession cities, states, and the entire United States will be looking for new ways to plug a massive budget deficit. Marijuana may be the only source of untapped tax revenues. And the need will be dire.

To put the U.S. debt levels in perspective, it took 205 years for the U.S. national debt to rise to $1 trillion. Yet the national debt rose by $1 trillion every month in the second quarter of 2020.[1] In order to keep the economy afloat during — and in the wake of — the COVID-19 pandemic and shutdown, stimulus has been piled on top of more stimulus as well as regular (and expanded) unemployment benefits. And the labor market figures are staggering.

As I write this book in August 2020, there are over 14.8 million unemployed Americans. In 1933, the worst year of the Great Depression, there were "only" 12.8 million. All of those unemployed people aren't paying taxes.[2]

Figure 22-1: Colorado Tax Revenues[3]

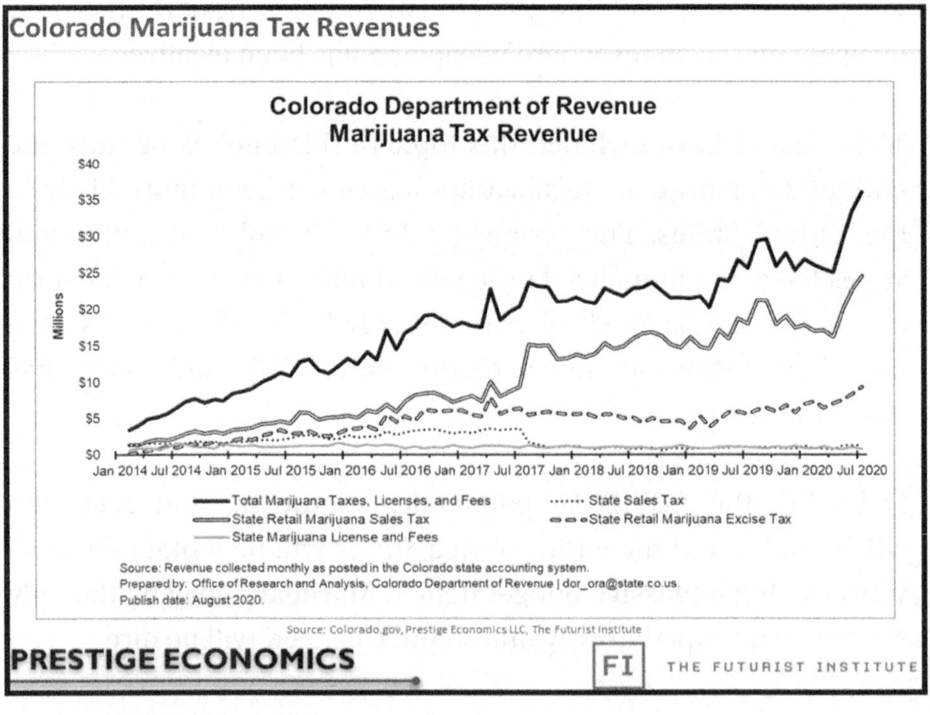

There are likely to be significant declines in revenue for federal income taxes, state income taxes, city income taxes (as in New York City), retail sales taxes, tourism taxes, franchise taxes, and property taxes in some markets. All of these shortfalls will put politicians on the path toward easy taxes to raise.

And the revenues in Colorado are compelling. In Figure 22-1, you can see how the marijuana tax revenues have risen in recent years. Furthermore, the cumulative revenue for Colorado between February 2014 and July 2020 was over $1.4 billion, as you can see in Figure 22-2. And Colorado is a small state with only 5.8 million people. Texas has around 5 times as many people.

Figure 22-2: Cumulative Tax Revenues[4]

Colorado Marijuana Tax Revenues

Marijuana Taxes, Licenses, and Fee Revenue

Calendar Year	Revenue By Calendar Year	Total Revenue Since Feb 2014
2014 (Feb - Dec)	$67,594,323	$67,597,323
2015	$130,411,173	$198,005,496
2016	$193,604,810	$391,610,306
2017	$247,368,473	$638,978,779
2018	$266,529,637	$905,508,416
2019	$302,458,426	$1,207,966,842
2020 (Jan - Jul)	$203,307,928	$1,411,274,770

Source: Colorado.gov, Prestige Economics LLC, The Futurist Institute

PRESTIGE ECONOMICS FI THE FUTURIST INSTITUTE

Not all states have been as successful with raising tax revenues on marijuana as Colorado. But the example is one that is likely to be heavily invoked as states approach 2021 with an impoverished mindset, hungry for low-hanging fruit — or weed, as it were. Plus, this isn't a small business.

In 2019, there were $12.4 billion in legal wages paid nationally to 340,000 people working cannabis jobs across the United States, with a total U.S. legal-market demand forecast of $13.6 billion. By 2025, the U.S. legal market is forecasted to have 743,000 jobs, with a legal-market demand size that would also be double the 2019 figure.[5]

The U.S. marijuana legal market in 2025 is likely to be much bigger market than lab-grown meat or hydroponics. And in the wake of COVID-19, the financial opportunities of marijuana tax revenues are likely to be much too large for economically-ailing governments to pass up.

LOOKING AHEAD

CHAPTER 23

FUTURE SCENARIOS

There are many ways that the future can play out. Part of being a futurist involves considering different scenarios for the future in order to foster discourse and consider several alternatives as well as identify the levers that could result in one future outcome over another. In this section, I will also share what I believe to be the most likely scenario.

Although some futurists might take umbrage at my preference for identifying what I consider to be the most likely scenario for the future, it is an important and practical part of my work as an advisor to companies, institutional investors, NGOs, and government entities to share what I asses to be the most likely scenarios.

Scenario: Technological Upside

Technology and innovation have tremendous potential to resolve some of the biggest challenges for the future of food and agriculture. In this scenario, calories will be provided for by new, emerging, and future innovations.

Ample water supplies for agricultural and industrial uses are made available with ease and other environmental ills of the world get resolved with technological solutions. This includes reducing emissions and reversing climate change.

In this scenario, GMOs and other technologies can be expected to drive global yields higher. Regenerative technology that can reduce global emissions has a material impact and food deserts are eradicated with the widespread creation of urban and suburban food facilities that include hydroponic, lab-grown meat, and other technologies. This utopian scenario is the least likely scenario in my opinion. It depends on a lot of technology being developed rapidly and deployed effectively.

Scenario: Breakdown and Bifurcation

In this scenario, technology and innovation cannot help people solve food and agricultural problems that accompany a rising population and global wealth. The demands and pressures on the planet are just too great. The needs of the global citizenry cannot be met equally — and disparities widen sharply. There are massive and extensive droughts and famines as water shortages become frequent and depleted reservoirs become increasingly polluted with pesticides. Global emissions rise sharply as well.

Geopolitically, a handful of countries keep a tight grip on fresh water and resource wars are frequent, pervasive, and brutal. Meanwhile, most of the world is a giant food desert. Bifurcated food dynamics make "clean food" and "real food" the hallmark of the ultrawealthy, while the vast majority of the populace eats substandard food and meat that is of generally poor quality.

This scenario does not seem like the most likely as seen from August 2020. However, this seems to be a more plausible scenario than the technological upside.

Scenario: Muddling Through

Throughout this book, I have tried to share a view of the potential trends and technologies that will impact the future, including the upside opportunities — as well as some downside risks. The risks largely stem from an expectation of rising population and limited water, land, and energy resources, as well as concerns about emissions and global conflict. Balanced against these challenges and risks, are opportunities presented by technologies and innovation that could improve the future.

But while neither a purely utopian nor dystopian outcome seems likely to come to fruition, an outcome in the middle seems likely. How good or bad the outcomes actually are will hinge on a number of factors, including the following:

1.) **The rate of population growth in the next 30 years.** Fewer people would take pressure off the system, while more people would stress global ag sharply. Population growth rates have slowed sharply in recent years, and COVID-19 could even create a baby bust. If it accelerates, such dynamics could reduce future global calorie needs.[1]

2.) **Support for initiatives that encourage sustainability.** This could come from NGOs, the government, or corporations. More support and more funding would likely lead to better outcomes.

3.) **Technological adoption, deployment, and efficiency.** Some technologies could greatly improve crop yields, provide for protein calories, and improve the environment. But funding, ROI, effective management, and incentives could be critical for successful adoption and use of these technologies.

4.) **Geopolitical diplomacy and trade.** If the Second Cold War happens, proxy wars could morph into resource wars. If global power competition cools off, proxy wars and resource wars are less likely. Continued solid levels of global trade would support global interdependence and could foster cooperation. Positive international relations could also foster shared or traded water resources internationally.

5.) **Calorie Consumption.** If emerging middle classes and the rest of the developing world aspire to mirror U.S. and industrialized nation total calorie and meat calorie consumption, stressors in global ag will be greater. If, however, preferences for total calorie and meat calorie consumption fall from current levels or do not rise to the American or industrialized levels, there would be significantly less pressure on global agriculture.

In short, the main levers of **people, priorities, technology, trade, and calorie consumption** will dictate the future of the decade ahead and beyond. Unfortunately, as seen from August 2020, dysfunctional governments, great power competition, and a Rousseauian world of limited resources may struggle to capture all of the upside potential ahead. But some potential will be captured. And while there is a potential for utopian or dystopian futures, we will likely land somewhere in the middle, although it will depend heavily on these five factors.

CHAPTER 24

PULLING EVERYTHING TOGETHER

My goal in writing this book was to examine, identify, and share my perspectives on the most important trends and technologies likely to disrupt and impact food and agriculture after the COVID-19 pandemic, shutdown, and recession. I hope that the big risks and opportunities are clear. The need to feed and nourish a growing global citizenry, while balancing the environmental limitations and impacts will be no easy task. Meanwhile, supply chain, security, and national security challenges could play out in the background if the world turns more darkly Rousseauian when water, food, and ag resource war risks increase.

Large numbers are often tough to grasp, but the data seem quite clear on these points. Demands on agribusiness and the planet will be high as the effort to keep calories, the environment, and geopolitics in balance become nothing short of a Herculean task. But we are not without hope. General purpose technologies like blockchain, quantum computing, IoT data, and automation could help make the world of food and agriculture more efficient, safer, and more secure.

The future of agriculture will also be greatly impacted by an industry-wide push into technologies like hydroponics and lab-grown meat. Substitution effects and efficiency gains exist, and there are opportunities to push crop yields higher. Meanwhile, ESG and sustainability mandates are likely to become increasingly important to maximize global production, while also addressing environmental risks and improving food deserts.

Further Learning

If you've enjoyed this book and you want to learn more about planning strategically for an uncertain future, the Certified Futurist and Long-Term Analyst™ — FLTA™ — training program at The Futurist Institute helps leaders incorporate trend and technology risks and opportunities into long-term strategic planning.

The FLTA™ has six distinct professional tracks, including consulting, national security, financial planning, accounting, legal, and standard professional. Plus, The Futurist Institute is accredited as a continuing education provider by the Certified Financial Planner Board of Standards®. Details about the FLTA™ and The Futurist Institute are online at www.futuristinstitute.org.

Your Next Steps

The future of food and agriculture will bring opportunities, risks, and challenges. But now that you have been informed about what big dynamics, trends, and technologies will impact the world we live in, you are better prepared for *The Future of Food and Agriculture After COVID*.

- Jason Schenker, August 2020

ENDNOTES

Chapter 1

1. Iredale, Will and Grimston, J. (10 October 2004.) Britain "Four Meals Away from Anarchy." *The Sunday Times*. Retrieved on 25 August 2020 from https://www.thetimes.co.uk/article/britain-four-meals-away-from-anarchy-fc9kfgc0w92

Chapter 2

1. NBER, FRED, World Bank, Prestige Economics. Retrieved February 17, 2017:
http://www.nber.org/chapters/c1567.pdf
https://fraser.stlouisfed.org/files/docs/publications/frbslreview/rev_stls_198706.pdf
http://databank.worldbank.org/data/reports.aspx?source=world-development-indicators#
2. Employment Indicators. Food and Agriculture Organization of the United Nations. Retrieved on 20 August 2020 from http://www.fao.org/faostat/en/#data/OE.
3. Ibid.
4. Ibid.
5. Employment in Agriculture. World Bank. Retrieved on 24 August 2020 from https://data.worldbank.org/indicator/SL.AGR.EMPL.ZS?locations=US-CN-KR-VN-1W.
6. Ibid.
7. Ibid.

Chapter 3

1. Land Use. Food and Agriculture Organization of the United Nations. Retrieved on 20 August 2020 from http://www.fao.org/faostat/en/#data/RL.
2. Ibid.
3. Ibid.

Chapter 4

1. Crops. Food and Agriculture Organization of the United Nations. Retrieved on 20 August 2020 from http://www.fao.org/faostat/en/#data/QC.

Chapter 5

1. Population Estimates and Projections. The World Bank. Retrieved on 10 May 2019 from https://datacatalog.worldbank.org/dataset/population-estimates-and-projections.
2. World Oil Outlook 2018. OPEC. Retrieved on 9 May 2019 from https://woo.opec.org/chapter.html?chapterNr=1&chartID=9. All OPEC data and graphs provided with permission. We very much appreciate the rights to use these.
3."What the World Eats." National Geographic. Retrieved on 20 August 2020 from https://www.nationalgeographic.com/what-the-world-eats/.
4. "Global and regional food consumption patters and trends." World Health Organization. Retrieved on 20 August 2020 from https://www.who.int/nutrition/topics/3_foodconsumption/en/.

Chapter 6
1. "What the World Eats." National Geographic. Retrieved on 20 August 2020 from https://www.nationalgeographic.com/what-the-world-eats/.
2. Ibid.
3. Ritchie, Hannah. (2017.) "Meat and Dairy Production." Our World In Data. Retrieved on 20 August 2020 from https://ourworldindata.org/meat-production.
Livestock Primary. Food and Agriculture Organization of the United Nations. Retrieved on 20 August 2020 from http://www.fao.org/faostat/en/#data/QL.
4. Ibid.
5. Ibid.
6. Ritchie, Hannah. (2020.) "Environmental impacts of food production." Our World In Data. Retrieved on 20 August 2020 from https://ourworldindata.org/environmental-impacts-of-food.
Poore, J., & Nemecek, T. (2018.) Reducing food's environmental impacts through producers and consumers. Science, 360(6392), 987-992. Retrieved on 20 August 2020 from https://science.sciencemag.org/content/360/6392/987.
7. Ritchie, Hannah. (2017.) "Meat and Dairy Production." Our World In Data. Retrieved on 20 August 2020 from https://ourworldindata.org/meat-production.
Alexander, P., Brown, C., Arneth, A., Finnigan, J., & Rounsevell, M. D. (2016). "Human appropriation of land for food: the role of diet." Global Environmental Change, 41, 88-98. Retrieved on 20 August 2020 from http://www.sciencedirect.com/science/article/pii/S0959378016302370?via%3Dihub#bib0330.
8. Ritchie, Hannah. (2020.) "Environmental impacts of food production." Our World In Data. Retrieved on 20 August 2020 from https://ourworldindata.org/environmental-impacts-of-food.
Poore, J., & Nemecek, T. (2018.) Reducing food's environmental impacts through producers and consumers. Science, 360(6392), 987-992. Retrieved on 20 August 2020 from https://science.sciencemag.org/content/360/6392/987.

Chapter 7

1. "Cultured Meat Market." Markets and Markets. Retrieved on 26 August 2020 from https://www.marketsandmarkets.com/Market-Reports/cultured-meat-market-204524444.html.
2. Kateman, Brian. (17 February 2020.) "Will Cultured Meat Soon Be A Common Sight In Supermarkets Across The Globe?" Forbes. Retrieved on 26 August 2020 from https://www.forbes.com/sites/briankateman/2020/02/17/will-cultured-meat-soon-be-a-common-sight-in-supermarkets-across-the-globe/#2b71a3097c66.
3. Mitchell, Natasha. (30 November 2017.) "Lab-grown burger with the lot: can we make meat more humane, and less polluting?" ABC Science. Retrieved on 26 August 2020 https://www.abc.net.au/news/science/2017-12-01/lab-made-meat-can-food-be-more-humane-and-less-polluting/9206542.

Chapter 8

1. "Hydroponics Market." Markets and Markets. Retrieved on 26 August 2020 from https://www.marketsandmarkets.com/Market-Reports/hydroponic-market-94055021.html.

Chapter 9

1. U.S. Census Bureau, E-Commerce Retail Sales [ECOMSA], retrieved from FRED, Federal Reserve Bank of St. Louis; https://fred.stlouisfed.org/series/ECOMSA, August 25, 2020.
2. U.S. Census Bureau, E-Commerce Retail Sales as a Percent of Total Sales [ECOMPCTSA], retrieved from FRED, Federal Reserve Bank of St. Louis; https://fred.stlouisfed.org/series/ECOMPCTSA, August 25, 2020.

Chapter 10

1. Speech by Darryl Willis, VP Oil, Gas, and Energy, Google Cloud at Google. Speech given at D2:Upheaval on 10 October 2018.
2. *Investopedia*. "Moore's Law." Retrieved on 7 November 2018 from https://www.investopedia.com/terms/m/mooreslaw.asp.
3. Gribbin, J. *Computing with Quantum Cats: From Colossus to Qubits*. Prometheus Books: New York. p. 92.
4. Jurvetson, S. (10 December 2016.) "Moore's Law Over 120 Years." Flickr. Retrieved on 1 November 2018 from https://www.flickr.com/photos/jurvetson/31409423572/.
5. Speech by Deepu Talla, Vice President and General Manager of Autonomous Machines, NVIDIA, at RoboBusiness on 26 September 2018.

Chapter 11

1. Farley, A. (2018.) "Bearer Bonds: From Popular to Prohibited" Investopedia. Retrieved on 24 August 2018 from https://www.investopedia.com/articles/bonds/08/bearer-bond.asp.
2. Image provided courtesy of Heritage Auctions, HA.com. Retrieved on 24 August 2018 from https://currency.ha.com/itm/miscellaneous/other/-1-000-000-us-treasury-bearer-bond/a/364-15547.s.
3. Lagarde, C. (26 April 2018.) "Statement by IMF Managing Director Christine Lagarde on Her Participation in the Paris Conference on Terrorism Financing." International Monetary Fund. Retrieved on 24 August 2018 from https://www.imf.org/en/News/Articles/2018/04/26/pr18150-lagarde-on-her-participation-in-the-paris-conference-on-terrorism-financing.
4. "Annual Economics Report." (June 2018.) Bank of International Settlements, p. 91-141. Retrieved from https://www.bis.org/publ/arpdf/ar2018e.pdf.
5. Strohm, Chris. (13 August 2020.) "U.S. Says It Seized Cryptocurrency From Three Terrorist Groups." Bloomberg. Retrieved on 15 August 2020 from https://www.bloomberg.com/news/articles/2020-08-13/u-s-says-it-seized-cryptocurrency-from-three-terrorist-groups?sref=GNzrTNIg .
6. Powell, J. "Semiannual Monetary Policy Report to the Congress." U.S. Federal Reserve. Retrieved on 12 July 2019 from https://www.federalreserve.gov/newsevents/testimony/powell20190710a.htm.
7. Image at end of chapter sourced from Göll, H. (1876.) *Galerie der Meister in Wissenscheft und Kunst. Meister der Wissenschaft I. Die Weisen und Gelehrten des Altherthums*. Second Edition. Leipzig: Verlag von Otto Spamer. P. 395.

Chapter 13

1. Statista Research Department. (1 July 2019.) " Robotics in the food and beverages industry sales worldwide 2017-2021, by country." Statista. Retrieved on 20 August 2020 from https://www.statista.com/statistics/1018917/robotics-food-and-beverages-industry-sales-value-worldwide/.

Chapter 15

1. Welsh, H. (9 November 2018.) "Social, Environmental & Sustainable Governance Shareholder Proposals in 2018." *Securities and Exchange Commission*, Sustainable Investments Institute. Retrieved on 12 July 2019 from www.sec.gov/comments/4-725/4725-4636528-176443.pdf.
2. Ibid.
3. "The Activist Investing Annual Review." (2020.) *Activist Insight.* Retrieved on 12 August 2020 from https://www.activistinsight.com/research/TheActivistInvesting_AnnualReview_2020.pdf.
4. "Shareholder Activism in 2019." (2020.) *Activist Insight.* Retrieved on 12 August 2020 from https://www.activistinsight.com/research/Shareholder_Activism_IN_2019.pdf.

Chapter 16

1. Gelski, Jeff. (20 May 2019.) "U.S. annual organic food sales near $48 billion." Food Business News. Retrieved on 20 August 2020 from https://www.foodbusinessnews.net/articles/13805-us-organic-food-sales-near-48-billion. See also Nunes, Keith. (10 June 2020.) "Organic food sales reach $50 billion in 2019." Food Business News. Retrieved on 20 August 2020 from https://www.foodbusinessnews.net/articles/16202-organic-food-sales-reach-50-billion-in-2019.
2. Ibid.
3. Silverstein, B. (15 October 2018.) "Within Five Years, Lab-Grown Meat May Open Up Kosher Options." The Canadian Jewish News. Retrieved on 16 September 2019 from https://www.cjnews.com/perspectives/big-ideas/within-five-years-lab-grown-meat-may-open-up-kosher-options. See also Castrodale, J. (23 March 2018.) "Rabbi Suggests That Lab Grown Pork Might Be Kosher." Vice. Retrieved on 16 September 2019 from https://www.vice.com/en_us/article/evqxdz/rabbi-suggests-that-lab-grown-pork-might-be-kosher. See also JTAStaff. (22 March 2018.) "Rabbi: Lab-grown pork could be kosher for Jews to eat – with milk." The Times of Israel. Retrieved on 16 September 2019 from https://www.timesofisrael.com/rabbi-meat-from-cloned-pig-could-be-eaten-by-jews-with-milk/.
4. "The Future of the Global Muslim Population." (27 January 2011.) Pew Research Center. Retrieved on 20 August 2020 from https://www.pewforum.org/2011/01/27/the-future-of-the-global-muslim-population/.
5. Shahbandeh, M. (23 October 2019.) "Market value of dairy milk alternatives worldwide in 2019, by category." Statista. Retrieved on 20 August 2020 from https://www.statista.com/statistics/693015/dairy-alternatives-global-sales-value-by-category/.
6. Richter, Felix. (17 June 2019.) "Alternative Meat Market Poised for Growth." Statista. Retrieved on 20 August 2020 from https://www.statista.com/chart/18394/meat-substitute-sales-in-selected-countries/.

Chapter 17

1. "Biomass Explained." Energy Information Agency. Retrieved on 9 May 2019 from https://www.eia.gov/energyexplained/index.php?page=biomass_waste_to_energy#tab1.
2. "Waste-to-Energy Electricity Generation Concentrated in Florida and Northeast." (8 April 2016). Energy Information Agency. Retrieved on 9 May 2019 from https://www.eia.gov/todayinenergy/detail.php?id=25732.
3. "Biomass Explained." Energy Information Agency. Retrieved on 9 May 2019 from https://www.eia.gov/energyexplained/index.php?page=biomass_waste_to_energy#tab2.
4. "Biomass Explained." Energy Information Agency. Retrieved on 9 May 2019 from https://www.eia.gov/energyexplained/index.php?page=biomass_waste_to_energy#tab2.

Chapter 18

1. Annual Energy Outlook 2019. (24 January 2019.) Energy Information Agency. Retrieved on 9 May 2019 from https://www.eia.gov/outlooks/aeo/tables_ref.php .
2. Ibid.
3. Ibid.
4. Ibid.

Chapter 19

1. Brodwin, Erin. (31 December 2017.) "Chocolate is on track to go extinct in 40 years." Business Insider. Retrieved on 26 August 2020 from https://www.businessinsider.com/when-chocolate-extinct-2017-12.
2. Davis, Aaron, Chadburn, Helen, Moat, Justin, O'Sullivan, Robert, and Hargreav, Serene (16 January 2019.) "High extinction risk for wild coffee species and implications for coffee sector sustainability." Science Advances, Vol. 5, no.1. Retrieved on 26 August from https://advances.sciencemag.org/content/5/1/eaav3473.
3. "Science and History of GMOs and Other Food Modification Processes." FDA. Retrieved on 26 August 2020 from https://www.fda.gov/food/agricultural-biotechnology/science-and-history-gmos-and-other-food-modification-processes.
4. "Are GMOs Safe to Eat?" (6 April 2016.) *WebMD*. Retrieved on 26 August 2020 from https://www.youtube.com/watch?v=Xnxt8KkIF8w.

Chapter 20

1. Drinking-water (14 June 2019.) World Health Organization. Retrieved on 26 August 2020 from https://www.who.int/news-room/fact-sheets/detail/drinking-water.
2. "The United Nations world water development report 2020: water and climate change." (2020). UNESCO World Water Assessment Programme. Retrieved on 26 August 2020 from https://unesdoc.unesco.org/ark:/48223/pf0000372985.locale=en. Pg. 22.
3. Ibid. Pg. 97.
4. Ibid.
5. Ibid. Pg. 56.

Chapter 21

1. Kenner, Bart and Jiang, Hui. (29 May 2020.) "Outlook for U.S. Agricultural Trade." United States Department of Agriculture. Retrieved on 20 August 2020 from https://downloads.usda.library.cornell.edu/usda-esmis/files/6m311p28w/fb494w31v/c534g826s/AES_112.pdf.
2. Ibid.

Chapter 22

1. U.S. Department of the Treasury. Fiscal Service, Federal Debt: Total Public Debt [GFDEBTN], retrieved from FRED, Federal Reserve Bank of St. Louis; https://fred.stlouisfed.org/series/GFDEBTN, July 21, 2020.
2. "Labor Force, Employment and Unemployment, 1929-39: Estimating Methods." U.S. Bureau of Labor Statistics, Retrieved on 22 July 2020 from: https://www.bls.gov/opub/mlr/1948/article/pdf/labor-force-employment-and-unemployment-1929-39-estimating-methods.pdf.
3."Marijuana Tax Data." Colorado Department of Revenue. Retrieved on 20 August 2020 from https://www.colorado.gov/pacific/revenue/colorado-marijuana-tax-data.
4. Ibid.
5. "Potential Cannabis Market Job Growth." CannaBit as cited by New Frontier Data. Retrieved on 26 August 2020 from https://newfrontierdata.com/cannabis-insights/potential-cannabis-market-job-growth/.

Chapter 23

1. Last, J. (2013.) *What to Expect, When No One's Expecting: America's Coming Demographic Disaster.* New York: Encounter Books, pp. 2-4. See also Coy, Peter. (29 July 2020). "Americans Aren't Making Babies and That's Bad for the Economy." Bloomberg. Retrieved on 15 August 2020 from https://www.bloomberg.com/news/articles/2020-07-29/coronavirus-pandemic-americans-aren-t-making-babies-in-crisis.

ABOUT THE AUTHOR

Mr. Schenker is the President of Prestige Economics and Chairman of The Futurist Institute. He has been ranked one of the most accurate financial forecasters and futurists in the world. Bloomberg News has ranked Mr. Schenker a top forecaster in 44 categories, including #1 in the world for his accuracy in 25 categories, including for his forecasts of the Euro, the British Pound, the Russian Ruble, the Chinese RMB, crude oil prices, natural gas prices, gold prices, industrial metals prices, agricultural commodity prices, and U.S. jobs.

Mr. Schenker was ranked one of the top 100 most influential financial advisors in the world by Investopedia in 2018. His work has been featured in *The Wall Street Journal*, *The New York Times*, and the *Frankfurter Allgemeine Zeitung*. He has appeared on CNBC, CNN, ABC, NBC, MSNBC, Fox, Fox Business, BNN, Bloomberg Germany, and the BBC. Mr. Schenker has also been a guest host of Bloomberg Television, and he is a contributor for *Bloomberg Opinion*.

Mr. Schenker attends OPEC and Fed events, and he has given keynotes for private companies, public corporations, industry groups, and the U.S. Federal Reserve. He has advised NATO and the U.S. government on the future of work, blockchain, Bitcoin, cryptocurrency, quantum computing, data analysis, forecasting, and fake news. Mr. Schenker has written 29 books. Twelve have been #1 bestsellers, including: *The Future After COVID, Jobs for Robots, Quantum: Computing Nouveau, Commodity Prices 101, Recession-Proof, Futureproof Supply Chain, Electing Recession, The Future of Finance is Now, The Future of Energy, The Dumpster Fire Election,* and *The Robot and Automation Almanac* for 2018 and 2020. Mr. Schenker also wrote *The Promise of Blockchain, Futureproof Supply Chain, The Fog of Data, Financial Risk Management Fundamentals, Midterm Economics, Spikes: Growth Hacking Leadership, Strategic Cost-Cutting, Strategic Cost-Cutting After COVID,* and *Reading the Economic Tea Leaves.* Mr. Schenker was featured as one of the world's foremost futurists in *After Shock.* His research was included in a June 2020 Pentagon report on global competition after COVID-19.

Mr. Schenker advises executives, industry groups, institutional investors, and central banks as the President of Prestige Economics. He also founded The Futurist Institute, for which he created a rigorous course of study that includes The Future of Work, The Future of Transportation, The Future of Data, The Future of Finance, Futurist Fundamentals, The Future of Energy, The Future of Leadership, The Future of Healthcare, The Future of Quantum Computing, and The Future After COVID. Mr. Schenker is also an instructor for LinkedIn Learning courses on Economic Indicators, Risk Management, Audit and Due Diligence, and Recession-Proof Strategies.

Mr. Schenker holds a Master's in Applied Economics from UNC Greensboro, a Master's in Negotiation, Conflict Resolution, and Peacebuilding from CSU Dominguez Hills, a Master's in Germanic Languages and Literature from UNC Chapel Hill, and a Bachelor's in History and German from The University of Virginia. He also holds a Certificate in FinTech from MIT, a Certificate in Supply Chain Management from MIT, a Certificate in Professional Development from UNC, a Certificate in Negotiation from Harvard Law School, a Certificate in Cybersecurity from Carnegie Mellon, and a Professional Certificate in Strategic Foresight from the University of Houston. Mr. Schenker holds the designations CMT® (Chartered Market Technician), and CFP® (Certified Financial Planner). He is also a Certified Futurist and Long-Term Analyst™ and holds the FLTA™ designation.

Before founding Prestige Economics, Mr. Schenker worked as a Risk Specialist at McKinsey and Company, where he provided content direction to trading, risk, and commodity project teams on six continents. Prior to McKinsey, Mr. Schenker was the Chief Energy and Commodity Economist at Wachovia, which is now Wells Fargo.

Based in Austin, Texas, Mr. Schenker is one of only 100 CEOs on the Texas Business Leadership Council, a non-partisan organization that advises Texas elected leadership at the state and federal level. Mr. Schenker is a Governance Fellow of the National Association of Corporate Directors. He also sits on multiple boards and is the VP of Finance on the Executive Committee of The Texas Lyceum, the preeminent non-partisan leadership group in Texas.

TOP FORECASTER ACCURACY RANKINGS

Prestige Economics has been recognized as the most accurate independent commodity and financial market research firm in the world. As the only forecaster for Prestige Economics, Jason Schenker is very proud that Bloomberg News has ranked him a top forecaster in 44 different categories since 2011, including #1 in the world in 25 different forecast categories.

Mr. Schenker has been top ranked as a forecaster of economic indicators, energy prices, metals prices, agricultural prices, and foreign exchange rates.

ECONOMICS TOP RANKINGS

#1 Non-Farm Payroll Forecaster in the World
#1 New Home Sales Forecaster in the World
#2 U.S. Unemployment Rate Forecaster in the World
#3 Durable Goods Orders Forecaster in the World
#6 Consumer Confidence Forecaster in the World
#7 ISM Manufacturing Index Forecaster in the World
#7 U.S. Housing Starts Forecaster in the World

ENERGY PRICE TOP RANKINGS

#1 WTI Crude Oil Price Forecaster in the World

#1 Brent Crude Oil Price Forecaster in the World

#1 Henry Hub Natural Gas Price Forecaster in the World

METALS PRICE TOP RANKINGS

#1 Gold Price Forecaster in the World

#1 Platinum Price Forecaster in the World

#1 Palladium Price Forecaster in the World

#1 Industrial Metals Price Forecaster in the World

#1 Copper Price Forecaster in the World

#1 Aluminum Price Forecaster in the World

#1 Nickel Price Forecaster in the World

#1 Tin Price Forecaster in the World

#1 Zinc Price Forecaster in the World

#2 Precious Metals Price Forecaster in the World

#2 Silver Price Forecaster in the World

#2 Lead Price Forecaster in the World

#2 Iron Ore Price Forecaster in the World

AGRICULTURAL PRICE TOP RANKINGS

#1 Coffee Price Forecaster in the World

#1 Cotton Price Forecaster in the World

#1 Sugar Price Forecaster in the World

#1 Soybean Price Forecaster in the World

FOREIGN EXCHANGE TOP RANKINGS

#1 Euro Forecaster in the World
#1 British Pound Forecaster in the World
#1 Swiss Franc Forecaster in the World
#1 Chinese RMB Forecaster in the World
#1 Russian Ruble Forecaster in the World
#1 Brazilian Real Forecaster in the World
#2 Turkish Lira Forecaster in the World
#3 Major Currency Forecaster in the World
#3 Canadian Dollar Forecaster in the World
#4 Japanese Yen Forecaster in the World
#5 Australian Dollar Forecaster in the World
#6 Polish Złoty Forecaster in the World
#7 Mexican Peso Forecaster in the World
#1 EURCHF Forecaster in the World
#2 EURJPY Forecaster in the World
#2 EURGBP Forecaster in the World
#2 EURRUB Forecaster in the World

More information about Prestige Economics:

www.prestigeeconomics.com

THE FUTURIST INSTITUTE

FI THE FUTURIST INSTITUTE

The Futurist Institute was founded in 2016 to help analysts, executives, and professionals incorporate technology and trend risks and opportunities into strategic planning. The Futurist Institute confers the Futurist and Long-Term Analyst™ (FLTA) designation and helps analysts become Certified Futurists™. Our courses have been approved for continuing education hours by the Certified Financial Planner Board of Standards (CFP Board).

Current Courses

The Future After COVID
The Future of Work
The Future of Data
The Future of Energy
The Future of Finance
The Future of Healthcare
The Future of Leadership
The Future of Transportation
Futurist Fundamentals
Quantum Computing
Write the Future

Visit The Futurist Institute:

www.futuristinstitute.org

PUBLISHER

Prestige Professional Publishing was founded in 2011 to produce insightful and timely professional reference books. We are registered with the Library of Congress.

Published Titles

Be the Shredder, Not the Shred
Commodity Prices 101
Electing Recession
Financial Risk Management Fundamentals
Futureproof Supply Chain
A Gentle Introduction to Audit and Due Diligence
Jobs for Robots
Midterm Economics
Quantum: Computing Nouveau
Reading the Economic Tea Leaves
Recession-Proof Career Strategies After COVID
Robot-Proof Yourself
Spikes: Growth Hacking Leadership
Strategic Cost-Cutting
Strategic Cost-Cutting After COVID
The Dumpster Fire Election
The Economics of Revolt and Revolution
The Fog of Data
The Future After COVID
The Future of Energy
The Future of Finance After COVID
The Future of Finance is Now

PUBLISHER

Published Titles

The Future of Food and Agriculture After COVID
The Promise of Blockchain
Write the Future
The Robot and Automation Almanac — 2018
The Robot and Automation Almanac — 2019
The Robot and Automation Almanac — 2020

Forthcoming Titles

Content Monster
Disruption Warfare
The Future of Healthcare
The Future of Travel and Leisure
The Robot and Automation Almanac — 2021

JOBS FOR ROBOTS

Jobs for Robots provides an in-depth look at the future of automation and robots, with a focus on the opportunities as well as the risks ahead. Job creation in coming years will be extremely strong for the kind of workers that do not require payroll taxes, healthcare, or vacation: robots. *Jobs for Robots* was published in February 2017. This book has been a #1 bestseller on Amazon.

FUTUREPROOF SUPPLY CHAIN

Futureproof Supply Chain examines more than a dozen major risks ahead — and how companies, industries, and individuals can prepare to be futureproof. The book was published in February 2019. This book has been a #1 bestseller on Amazon.

THE FUTURE AFTER COVID

In *The Future After COVID,* top-ranked forecaster and Chairman of The Futurist Institute Jason Schenker offers a futurist perspective into the potential long-term changes, challenges, and opportunities that COVID-19 is likely to have for over a dozen different critical fields and industries. The book was published in April 2020. This book has been an international bestseller.

DISCLAIMER

FROM THE AUTHOR

The following disclaimer applies to any content in this book:

This book is commentary intended for general information use only and is not investment advice. Jason Schenker does not make recommendations on any specific or general investments, investment types, asset classes, non-regulated markets, specific equities, bonds, or other investment vehicles. Jason Schenker does not guarantee the completeness or accuracy of analyses and statements in this book, nor does Jason Schenker assume any liability for any losses that may result from the reliance by any person or entity on this information. Opinions, forecasts, and information are subject to change without notice. This book does not represent a solicitation or offer of financial or advisory services or products; this book is only market commentary intended and written for general information use only. This book does not constitute investment advice. All links were correct and active at the time this book was published.

DISCLAIMER

FROM THE PUBLISHER

The following disclaimer applies to any content in this book:

This book is commentary intended for general information use only and is not investment advice. Prestige Professional Publishing, LLC does not make recommendations on any specific or general investments, investment types, asset classes, non-regulated markets, specific equities, bonds, or other investment vehicles. Prestige Professional Publishing, LLC does not guarantee the completeness or accuracy of analyses and statements in this book, nor does Prestige Professional Publishing, LLC assume any liability for any losses that may result from the reliance by any person or entity on this information. Opinions, forecasts, and information are subject to change without notice. This book does not represent a solicitation or offer of financial or advisory services or products; this book is only market commentary intended and written for general information use only. This book does not constitute investment advice. All links were correct and active at the time this book was published.

Prestige Professional Publishing, LLC

4412 City Park Road #4

Austin, Texas 78730

www.prestigeprofessionalpublishing.com

ISBN: 978-1-946197-66-5 *Paperback*

978-1-946197-67-2 *Ebook*

Made in United States
Orlando, FL
04 November 2026

90202253R00105